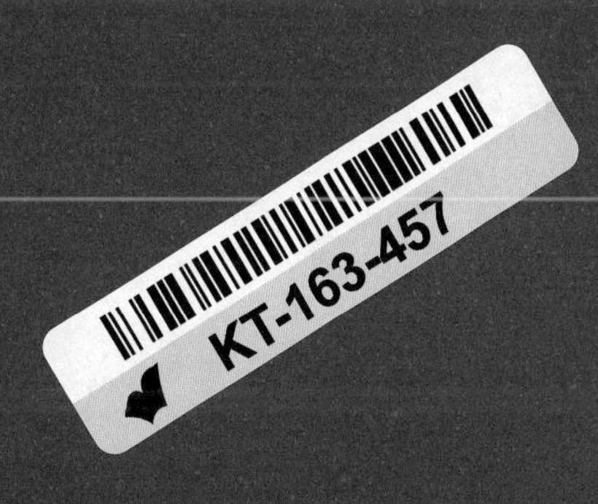
KT-163-457

FLASHPOINTS IN SCIENCE

Exploring the causes, effects and triggers of major 20th-century discoveries

Anne Rooney

An Hachette UK Company
www.hachette.co.uk

This edition published in 2016 by Bounty Books,
a division of Octopus Publishing Group Ltd
Carmelite House
50 Victoria Embankment
London, EC4Y 0DZ
www.octopusbooks.co.uk

First published in 2016 by Bounty Books, a division of Octopus Publishing Group Ltd

ISBN: 978-0-75372- 985-4

A CIP catalogue record for this book is available from the British Library

Acknowledgements
Publisher: Samantha Warrington
Design: The Urban Ant Ltd
Art Director: Miranda Snow
Editor: Phoebe Morgan
Picture Research: Liz Allen
Senior Production Manager: Peter Hunt

Printed and bound in Hong Kong

10 9 8 7 6 5 4 3 2 1

CONTENTS

CHAPTER ONE
1900–1919 4

CHAPTER TWO
1920–1939 38

CHAPTER THREE
1940–1959 72

CHAPTER FOUR
1960–1979 102

CHAPTER FIVE
1980–1999 144

$\frac{m u_i}{\sqrt{1-u^2}}$ Impuls
$m\left(\frac{1}{\sqrt{1-u^2}}-1\right)$ Kin Energ.
Hyp.
$x=\frac{x'+vt'}{\sqrt{1-v^2}}$ $y=y'$ $z=z'$

1900–1919

As the 20th century began, a sense of optimism and triumph dominated the scientific scene. The previous century had seen significant shifts in what was known and the possibilities of technology. Charles Darwin and Richard Owen had explained evolution; Charles Lyell had outlined the geological history of the Earth; psychology had emerged as a new science; and James Clerk Maxwell had identified the electromagnetic spectrum. Louis Pasteur had explained how microbes could cause disease and had made successful vaccines, while the development of anaesthetics and antiseptics made surgery viable. Mendeleev had published the periodic table of the elements, explaining the behaviour of chemicals. The discovery of dinosaur fossils had transformed the distant past. The final decade of the 1800s had seen the discovery of X-rays, the electron, viruses, antibodies and a treatment for diphtheria that represented the first systematic approach to disease. For young people entering science, there was even bewilderment about what they would do. What was left to discover?

It turned out that there was quite a lot still to discover. The first two decades of the 20th century saw huge strides, particularly in physics. This was the time of Einstein's greatest work, of the development of quantum theory and of understanding the structure of the atom. But the period was to end in catastrophe and slaughter, with the First World War and then the flu pandemic of 1918. It is hardly surprising that the greatest achievements of these two decades were crammed into the first 15 years.

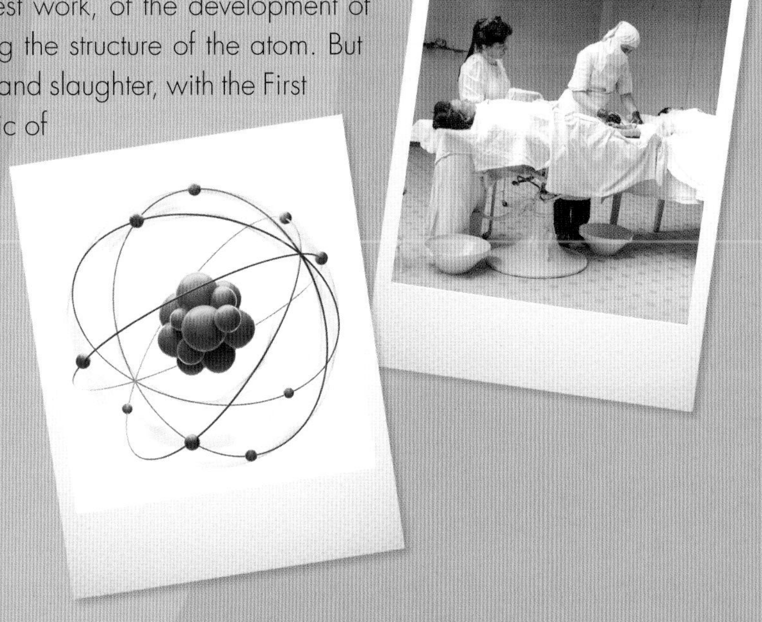

1900

QUANTUM THEORY: WAVES AND PARTICLES AS ONE

'Experience will prove whether this hypothesis is realized in nature.'

– Max Planck on the existence of energy quanta

Max Planck (right) with Danish physicist Niels Bohr; both were key figures in the development of quantum theory.

When the young Max Planck asked his professor, Philipp von Jolly, about the prospects of a career in physics, von Jolly advised him that physics was pretty much complete, with nothing left to discover. Fortunately, Planck ignored his advice. He went on to lay the foundations of quantum theory that, together with Einstein's theory of relativity, completely revolutionized physics – and has left us with plenty still to discover.

In 1900, Planck was investigating one of the few puzzles remaining in classical physics. A black body does not emit electromagnetic radiation in the way that would be expected. In physics, a 'black body' is an opaque, non-reflective object. If it is kept at a constant temperature, it emits electromagnetic radiation. The spectrum emitted depends on the temperature. Even though it looks black at room temperature, it is emitted infrared (which we can't see). As it heats up, it glows dull red, bright red through orange and eventually white. This so-called black-box radiation confounded all attempts to be shoehorned into the laws of physics. In what he later called an 'act of desperation', Planck stopped treating light as a wave and tried using discrete chunks of energy of fixed amounts in his equations. These chunks he called 'quanta' (singular, quantum). Suddenly, it all worked.

But everyone *knew* light was a wave – it had been known for centuries. Few people were enthusiastic about his suggestion (including Planck himself). Yet now, quanta of energy lie at the heart of physics and chemistry. They explain how chemicals react together, how electricity and electromagnetic radiation behave, and the structure of atoms.

FLASHPOINT FACT

One consequence of quantum theory is that everything has a wavelength – even you. Your wavelength is very small and will never have a noticeable effect; you will always behave like a particle.

TIMELINE

1679 Christiaan Huygens describes light as a wave of energy.

1704 Isaac Newton describes light as being made of 'corpuscles' – tiny, weightless particles.

1900 Max Planck suggests that energy comes in discrete chunks, or 'quanta'.

1905 Albert Einstein proposes the existence of the photon as a quantum of light energy.

1909 Geoffrey Taylor's double-slit experiment shows that light can behave as particles or as waves.

1913 Robert Millikan calculates the charge of an electron as 1.592×10^{-19} coulomb.

1913 Niels Bohr suggests a structure for the atom that takes account of quantum ideas.

FLASHPOINTS

1704
PARTICLE OR WAVE?

Two thousand years ago, the Roman philosopher Lucretius claimed that light is made of particles. Isaac Newton favoured the particle theory, too, in 1704, but by the time Newton was writing the wave theory promoted by Christian Huygens in 1679 was more popular. In 1800, Thomas Young shone light through an apparatus with two slits and found interference patterns on the far side. This supported the theory of light as waves. In the 1870s, James Clerk Maxwell showed visible light as part of the electromagnetic spectrum. The parts are distinguished by wavelength, so the nature of light seemed finally to have been determined.

1909
CANNY PHOTONS

In 1909, Geoffrey Taylor repeated Young's double-slit experiment, but firing only one photon at a time. There should be no interference pattern, as the photon must go through one slit or the other. Yet interference patterns did develop. Each photon appeared to split in half and go through both slits at the same time, then recombine into a single photon on the other side. When Taylor then put a photon detector at each slit, the pattern changed. Now the photons behaved only as particles. They seemed to 'know' that there was a particle detector present. They behaved as waves when he was looking for waves and particles when he was looking for particles.

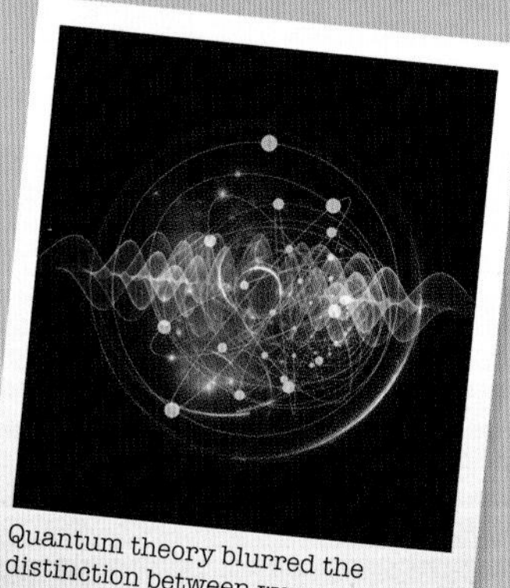

Quantum theory blurred the distinction between waves and particles.

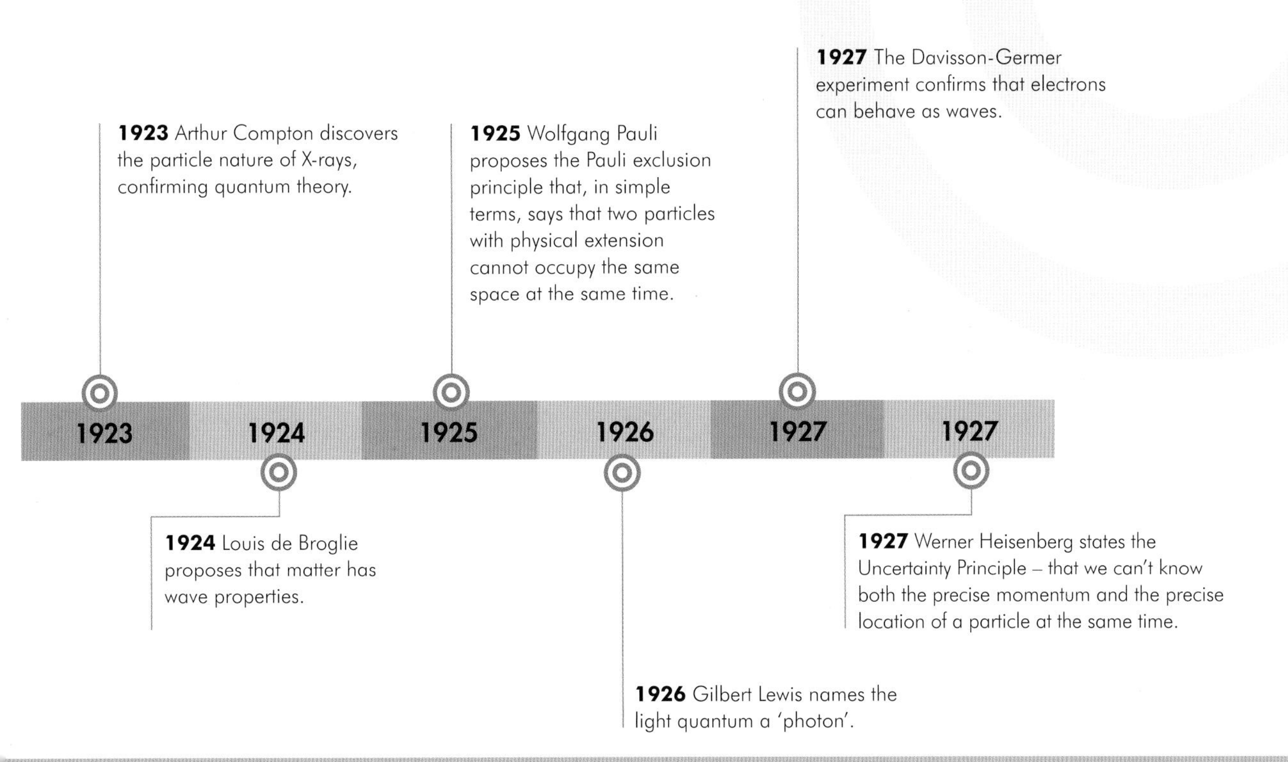

1923
MORE QUANTA

The more famous particle of energy is the electron, the first subatomic particle discovered. In 1909, Robert Millikan and Harvey Fletcher carried out an experiment in which they suspended a droplet of oil in an electric field, then let it fall and calculated the exact force needed to hold it in place against gravity. From this, they calculated the charge of an electron: 1.592×10^{-19} coulomb, the first measurable energy quantum. In 1923, Arthur Compton showed the quantum nature of X-rays. He showed that X-rays fired at a thin sheet of gold leaf bounce off with reduced energy (a larger wavelength), and the change in wavelength is always a fixed amount – a quantum.

1927
WAVES ALL AROUND

Just as light can act as a particle or a wave, so can other subatomic particles. In 1924, Louis de Broglie calculated the wavelength of a photon, deriving an equation that could be used to calculate the wavelength of other particles, too. That other particles can actually behave as waves was demonstrated in 1927 by Clinton Davisson and Lester Germer. They fired electrons at a nickel crystal and saw diffraction as though the electrons were waves. As soon as we think of matter as waves, any kind of certainty about its position disappears. Indeed, the Uncertainty Principle, stated by Werner Heisenberg in 1927, says just this: we can either know one set of properties with certainty or another, but never both.

1901

LANDSTEINER AND THE IDENTIFICATION OF BLOOD GROUPS

'The reactions follow a pattern, which is valid for the blood of all humans...Basically, in fact, there are four different types of human blood, the so-called blood groups.'

– Karl Landsteiner, in a lecture accepting his Nobel Prize in 1930 (translated by Pauline M.H. Mazumdar)

Landsteiner's discovery of blood groups made transfusions safer.

Surgery beyond simple amputations was made possible in the 19th century through the development of anaesthetics and antiseptics, but blood loss could still cause death. Early attempts to give blood transfusions usually failed. Karl Landsteiner found out why, making blood transfusions reliable and saving millions of lives over the last century.

Landsteiner investigated the effect of mixing animal and human blood, and often blood from two different human subjects. He found that it frequently led to the red blood cells clumping, then sometimes breaking apart, releasing haemoglobin into the blood. This is dangerous, and can be fatal. It did not always happen, though: sometimes he could mix blood from different people without the cells clumping.

Blood comprises red blood cells, white blood cells and plasma – a pale yellow fluid. Through his experiments, Landsteiner found in 1901 that it was something in the plasma that caused problems with the blood cells. If he mixed red blood cells with incompatible blood plasma, the clumping (agglutination) occurred. He worked out that blood falls into different groups, some of which are compatible and some not. He divided blood into groups A, B and C (now known as O). He discovered that blood of the same group does not cause problems, and group O (or C) can accept groups A and B – but A and B cannot accept O. Alfred von Decastello and Adriano Sturli discovered the fourth type, AB, in 1902. People with AB-group blood can accept any type of blood, but AB blood can only be given to AB patients.

The first successful blood transfusion followed in 1906, carried out by George Washington Crile in Cleveland, USA, using matched blood groups. Other distinctions between blood types have been found since Landsteiner's basic groups. More than 30 systems of blood grouping are now known.

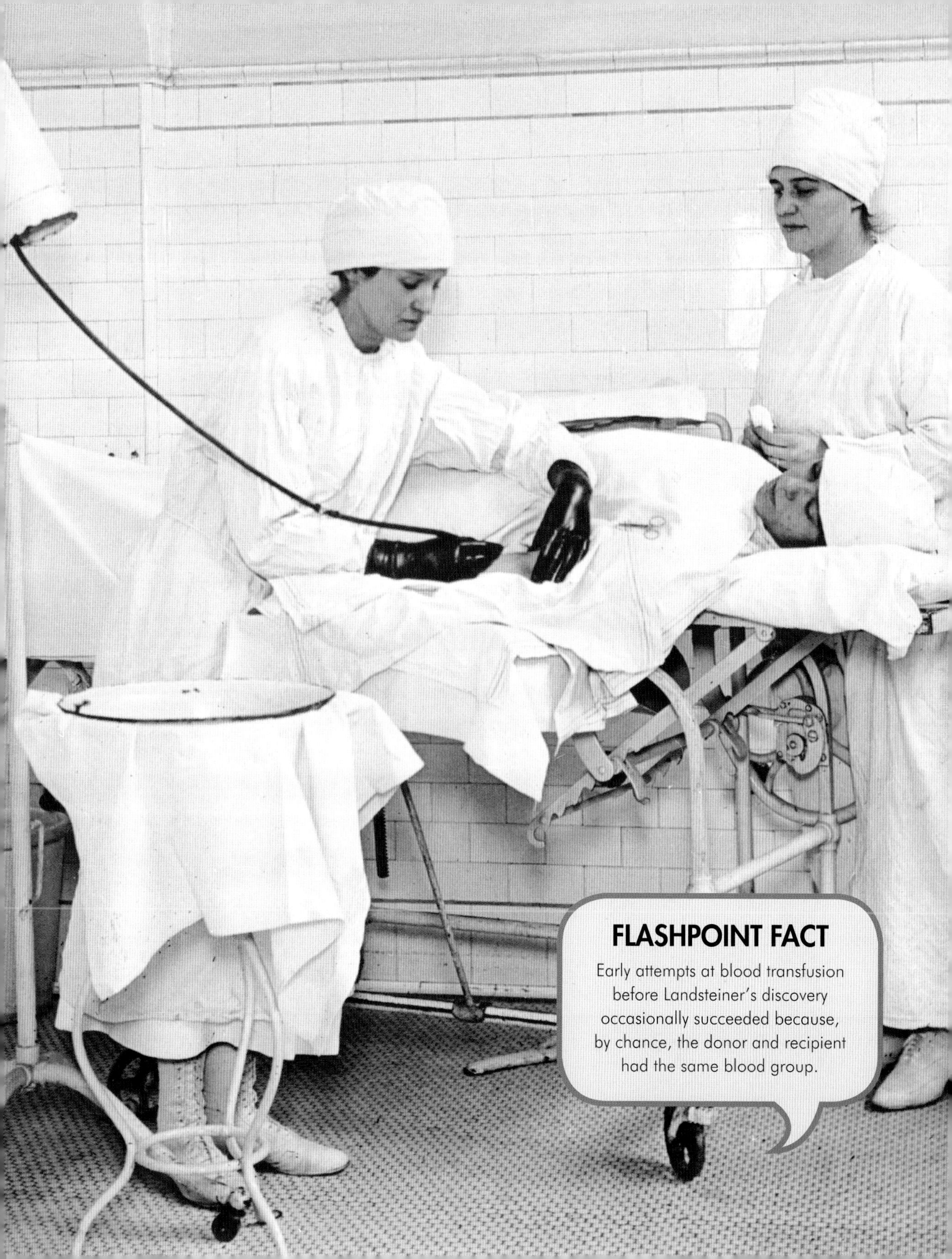

FLASHPOINT FACT

Early attempts at blood transfusion before Landsteiner's discovery occasionally succeeded because, by chance, the donor and recipient had the same blood group.

TIMELINE

1665 Richard Lower demonstrates blood transfusion between two living dogs to the Royal Society in London. Both survive.

1667 Jean-Baptiste Denys successfully transfuses blood from a sheep to a 15-year-old boy.

1668 The Royal Society in Britain, and the French government, ban further experiments with blood transfusion.

1670 The Vatican condemns blood transfusion.

1818 Dr James Blundell carries out the first successful human-to-human blood transfusion to treat bleeding after childbirth.

1840 Samuel Armstrong Lane, helped by Blundell, carries out the first whole transfusion to treat haemophilia.

1875 Leonard Landois reported that mixing blood from other animals with human blood caused blood cells to clump together and sometimes break up.

1901 Karl Landsteiner discovers that blood falls into different groups that are partially incompatible; he names these A, B and O.

1902 Alfred von Decastello and Adriano Sturli define an additional blood group, AB.

1665 | 1667 | 1668 | 1670 | 1818 | 1840 | 1875 | 1901 | 1902

FLASHPOINTS

1665
EARLY ATTEMPTS

The earliest attempts at blood transfusions moved blood between animals, and then between humans and animals. An infusion of animal blood can cause a fatal allergic reaction, but when only small amounts of blood were used this did not always happen. Dr Jean-Baptiste Denys successfully transfused sheep blood into two patients who survived, but two later patients, one transfused with blood from a calf, died. Attempts at transfusion stopped for around 150 years. Interest in it grew again in the early 19th century, with some successful human-to-human transfusions. It was risky, though, as it frequently led to problems.

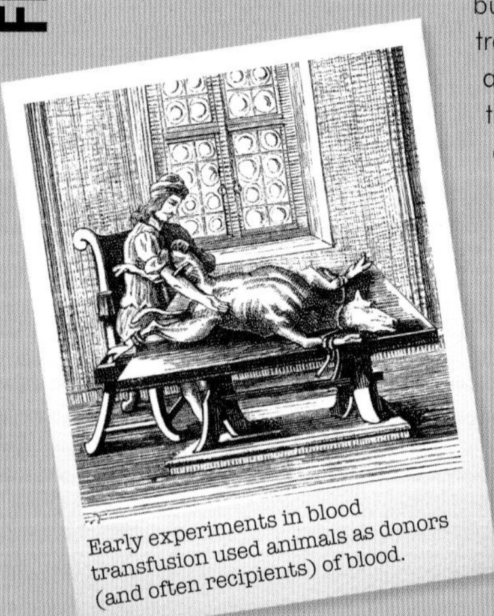

Early experiments in blood transfusion used animals as donors (and often recipients) of blood.

1901
THE SCIENCE BEHIND THE NEEDLE

Why should transfusions sometimes go well and sometimes be disastrous? This puzzled medical researchers until Landsteiner's discoveries revealed what was going on. If a person is given incompatible blood, their body's immune system attacks the new blood cells, which first form clumps and then disintegrate. The blood groups Landsteiner identified were A, which has A-antigens inside red blood cells and B-antibodies in the plasma; B, which has B-antigens in the cells and A-antibodies in the plasma; and C (now called O), which has no antigens in the cells and both types of antibodies in the plasma. AB, discovered later, has both types of antigens in the red blood cells and no antibodies. Antibodies are in the blood plasma (the liquid that carries the blood cells), which is no longer used in transfusions.

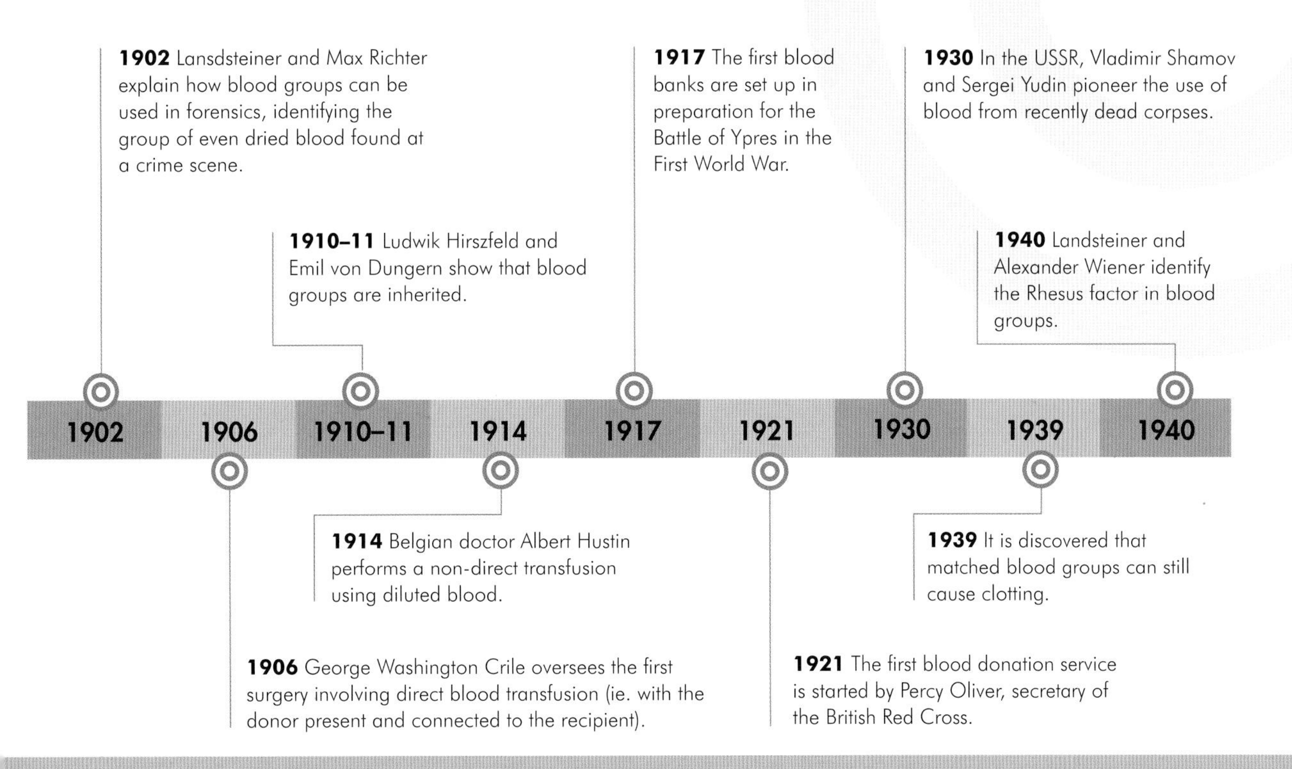

1910–11
BLOOD AND HEREDITY

Blood grouping has more uses than transfusion. In 1910–1, Ludwik Hirszfeld and Emil von Dungern discovered that ABO blood group is an inherited characteristic, and in 1924 Felix Bernstein demonstrated the slightly complex pattern of inheritance. This made it possible to use blood grouping to help establish – or at least to disprove – paternity. For example, two parents with group O can only have a group-O child. Two group-A parents can produce a child with group-A or group-O blood, and similarly two group-B parents can have a child with B or O blood type. On a larger scale, studying the blood groups of entire populations can give an indication of migration patterns in the past. This was discovered near the end of the First World War, when Hanna and Ludwig Hirszfeld found a sharp drop in A-type blood at the German border with Poland.

1914
BLOOD TRANSFUSIONS IN THE FIRST WORLD WAR

Understanding blood groups enabled blood transfusions and made possible more complicated surgery, including organ transplants. The first human-to-human transfusion was in 1906, and involved having the donor alongside the recipient. Transfusions of blood that had already been taken from the donor began in 1914 – just in time for the First World War, but blood could not be preserved long enough for transfusions to be widespread. By the end of the war, soldiers in some armies wore tags that gave their blood group, speeding up the process of finding compatible blood if they were injured. Blood could be kept for a short time using sodium citrate to prevent clotting. From 1930, the red blood cells were removed from the plasma and could be kept for longer.

1902

THE CHANGING FACE OF PAST LIFE: *T. REX*

'Quarry No. 1 contains [several bones] of a large carnivorous dinosaur not described by Marsh...I have never seen anything like it from the Cretaceous.'

– Barnum Brown, in a letter to the American Natural History Museum describing his find in 1902

Barnum Brown oversees the arrangements of bones from a *Tyrannosaurus rex*.

In 1900, the first hints of a thrilling monster from the past surfaced in Wyoming, USA. The eccentric palaeontologist Barnum Brown uncovered part of the skull and jaw, and some formidable teeth, from a dinosaur he named *Dynamosaurus imperiosus*. Two years later, in 1902, he found a complete skeleton. Not recognizing it as the same creature, he called the new, terrifyingly large carnivore *Tyrannosaurus rex*. It's difficult to imagine now how the discovery of *T. rex* must have changed the perception of dinosaurs, previously known mostly through the fish-eating plesiosaurs and huge land-going plant-eaters.

Barnum Brown, named after a circus strongman, was something of a real-life Indiana Jones. He attended digs in a full-length beaver-fur coat and a top hat, used dynamite to blast apart hillsides and did a great deal to popularize dinosaurs. He was also a spy. He was one of the greatest dinosaur hunters, finding so many fossils that there are still unopened crates of his finds awaiting investigation.

T. rex has become the public ambassador of the dinosaurs, more familiar than many living creatures. The iconic killing machine has captivated generations and ensured continuing investment in dinosaur palaeontology.

T.rex is the only dinosaur that is commonly known to the public by its full scientific name, described by American palaeontologist Robert T. Baker as 'just irresistible to the tongue.'

FLASHPOINT FACT

A little bit of a *T. rex* was found in 1892 and the dinosaur named *Manospondylus gigas*. It was proved to be the same as *T. rex* in 2000, and the name of *T. rex* should then have been changed – but that was never going to work!

TIMELINE

1677 Robert Plot includes the first illustration of a dinosaur bone in a book, *The Natural History of Oxfordshire*. He believed it to be a thigh bone from a giant human.

1808 French naturalist Georges Cuvier describes a giant marine lizard from a fossil found in Maastricht, Holland, and seized by Napoleon's army as a war trophy.

1809 William Smith finds an *Iguanodon* shin bone, but it was not recognized as such until the 1970s.

1811 Mary Anning finds a large ichthyosaur fossil at Lyme Regis, England.

1824 William Buckland names *Megalosaurus* from a jawbone that had been in Oxford's university museum since 1818; this was the first named dinosaur.

1825 Gideon Mantell describes *Iguanodon*.

1842 Richard Owen names a new sub-order, Dinosauria.

1677 | 1808 | 1809 | 1811 | 1824 | 1825 | 1842

FLASHPOINTS

1808
LONG AGO AND FAR AWAY

People have been finding and digging up the fossilized bones of long-dead animals for centuries without knowing what they were. In China, they were often considered to be dragon bones. The first realization that fossils were the remains of prehistoric creatures came with the work of Georges Cuvier in Napoleon's France. In 1808, he identified a war trophy taken from Maastricht in Holland as the fossil of an extinct marine lizard, now called *Mososaurus*. The recognition that there had been former ages, with creatures now extinct, was revolutionary. In the following years, fossil hunters such as Mary Anning and Gideon Mantell uncovered evidence of large creatures that went by sea and by land.

Mososaurus, uncovered in a cave in Maastricht, Netherlands, was misidentified as both a crocodile and a sperm whale.

1842
THE DAWN OF THE DINOSAURS

In 1842, Richard Owen put the pieces together, establishing the new sub-order (now clade) Dinosauria to hold these 'fearfully great reptiles'. The suggestion that there had been an age long before humans when the Earth was ruled by giant, egg-laying lizards was a challenge to the Christian view that God had created all creatures in an unchanging world 6,000 years previously.

The second half of the 19th century was the age of the great dinosaur hunters. Othniel Marsh and Edward Cope competed to find the most and biggest dinosaurs, uncovering *Stegosaurus*, *Triceratops*, *Allosaurus* and *Diplodocus* between the 1870s and 1900. Cope alone named 1,200 species of dinosaur.

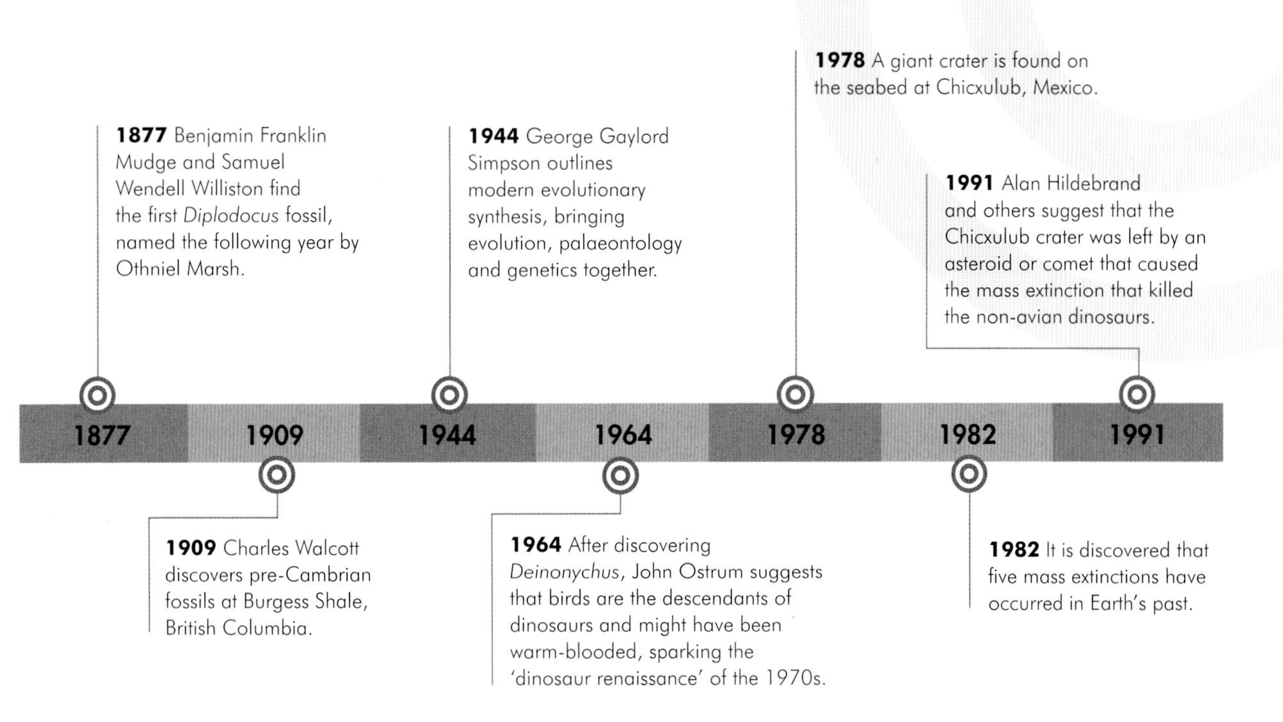

1909
THE BURGESS SHALE

Then in 1909 came another revolutionary discovery: a vast field of fossils, far older than those of the dinosaurs. At Burgess Shale in Canada, Charles Walcott discovered soft-bodied animals and plants fossilized in great numbers. He returned year after year, amassing over 65,000 fossils, but their full importance was not recognized until the 1960s. The fossils of the Burgess Shale date from the Cambrian explosion, the sudden and rapid diversification of life around 542 million years ago. The fossils are also now studied by climatologists trying to project the effect of warming on Earth.

The Burgess Shale contains millions of fossils of trilobites. They were among the first discovered.

1940s
FOSSILS AND EVOLUTION

Marrying the fossil record with the way evolution was thought to progress was a challenge. That was accomplished in the 1930s and 1940s by several biologists, geneticists and palaeontologists. The modern evolutionary synthesis developed in the 1940s remains, in essence, the main paradigm in evolutionary theory. It shows evolution to be gradual, not directed towards any goal or perfection, and with many unsuccessful organisms failing along the way. It also demonstrates how the microevolution of genetics can produce the macroevolution seen in the fossil record.

1902

DRIBBLE AND DROOL: PAVLOV'S HUNGRY DOGS

'Don't become a mere recorder of facts, but try to penetrate the mystery of their origin.'

– Ivan Pavlov

A researcher working with dogs in Pavlov's laboratory in St Petersburg, Russia.

Pavlov's dribbling dogs are familiar to everyone, and especially dog owners. Following experiments on digestion in the 1890s, the Russian physiologist Ivan Pavlov turned his attention to the link between physiology and psychology in creating a reflex. A reflex is an action the body performs automatically in response to a stimulus, without any conscious intention. Snatching your hand away from a hotplate is a reflex, as is crying out at pain.

Pavlov began by exposing dogs to a stimulus, such as a ringing bell or flashing light, when feeding them. By repeating this several times, he led the dogs to associate the secondary stimulus with food. The dogs' natural reflex to produce saliva at the smell of food was soon transferred to the secondary stimulus (the sound or light he had associated with food). Continuing his experiments, Pavlov found that if he produced the conditioned stimulus (the bell, light or other stimulus) without giving food, the association would eventually break and the dogs would no longer salivate in response to it – so the conditioned response would become 'extinct'. It was easy to reassert it, though, by reconnecting the two.

Pavlov's discovery – that dogs would produce saliva in response to a stimulus they had come to associate with food – is now called classical conditioning. His work was the first to show a connection between physiology and psychology. It set the stage for the investigation of learning mechanisms in the 20th century.

FLASHPOINT FACT

Pavlov also used whistles at different pitches with his dogs, leading them to associate a particular pitch with food and drool only on hearing a whistle of the right pitch.

TIMELINE

1879 Wilhelm Wundt opens the first experimental psychology laboratory, in Leipzig, Germany.

1885 Pavlov begins his research on digestion and reflexes.

1897 Pavlov publishes his first findings on reflexive production of saliva.

1902 Pavlov publishes further findings on digestive reflexes and classical conditioning.

1904 Pavlov wins the Nobel Prize for Medicine or Physiology for his work on classical conditioning.

1913 The Behaviourist school in psychology is established by John B. Watson to focus on the observable effects (behaviour) of mental processes.

1879 | 1885 | 1897 | 1902 | 1904 | 1913

FLASHPOINTS

1879
BEHAVIOURISM AND BEFORE

The subject of psychology emerged in Germany, developed first by Wilhelm Wundt in the 1870s. He was interested in revealing the structure of the mind. His experimental approach involved exposing human subjects to stimuli and asking them to report their responses. The subject's active role in reporting led to criticisms that the approach was unreliable and subjective. The Behaviourist school, established by John B. Watson in 1913, reacted against these limitations by focusing only on measurable behaviour and physiological response. Pavlov's work – biological as much as psychological – fit this model and was hugely influential. Many of the behaviourists worked extensively with animals to exclude any polluting involvement of the subject's consciousness.

1920–4
LITTLE ALBERT AND LITTLE PETER

In one particularly famous and unethical experiment, Watson conditioned a fear response in a toddler, known as Little Albert. 'Albert' was initially unafraid of a white rat, but Watson built up an association in Albert's mind between the rat and frightening noises. Eventually, the child came to fear the rat and back away from it, demonstrating that classical conditioning works in humans. In 1923–4 Mary Cover Jones put it to better use. She worked with a child called Peter who was afraid of white rabbits. She introduced a rabbit with a pleasant stimulus (food) and then brought the rabbit slowly closer. Through 'direct conditioning' she cured his fear. Controlled, gradual exposure is now the basis of phobia treatment, called desensitization.

Ivan Pavlov.

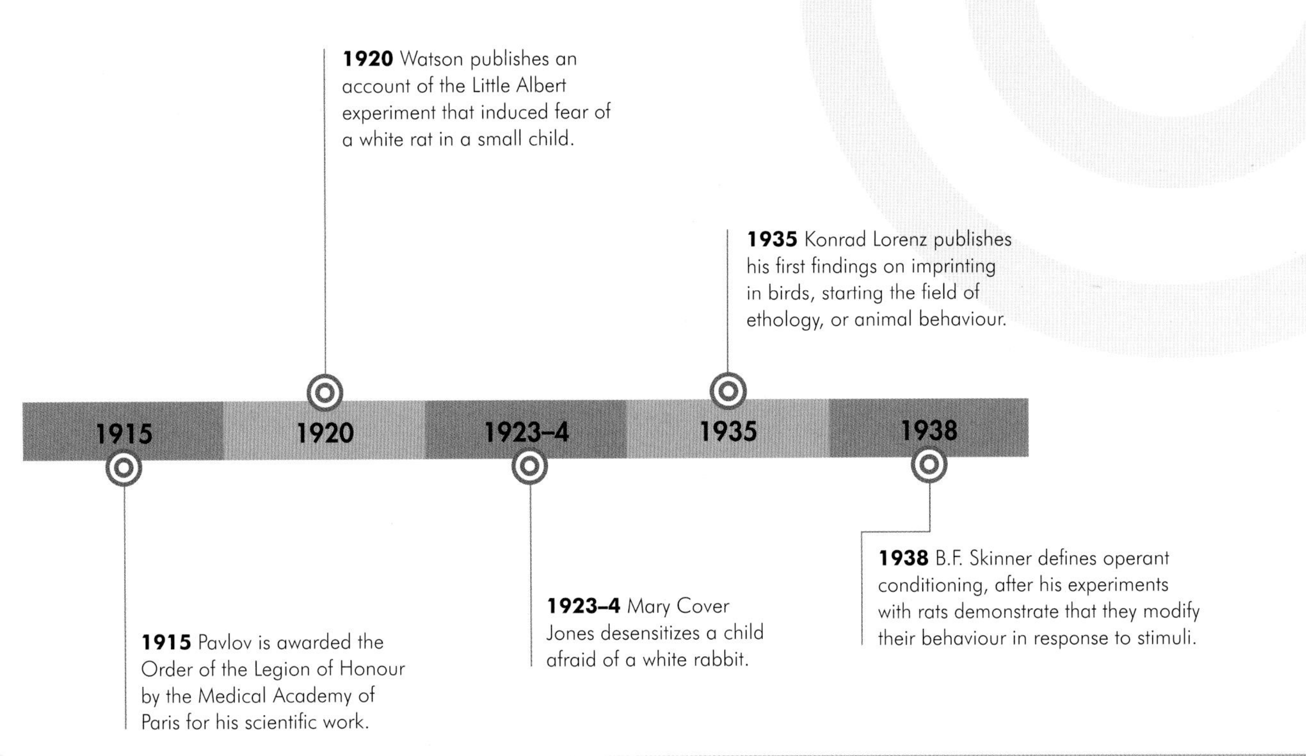

1930s
RATS IN A BOX

In the 1930s, psychologist B.F. Skinner worked with rats in a box that became called a 'Skinner box'. He trained rats to repeat behaviour in response to receiving a reward, such as sugar water. He also used negative stimulus (a mild electric shock) to condition rats' behaviour. This is called operant conditioning. While classical conditioning produces an involuntary response, operant conditioning changes. Skinner was a committed behaviourist, the approach that dominated psychology for much of the 20th century. In 1971, he argued that free will is an illusion and all our actions are the result of conditioning and instincts.

1935
BEST INSTINCTS

Konrad Lorenz carried out experiments with graylag geese in the 1930s. He found that chicks are biologically primed to 'imprint' on a moving object as soon as they hatch. Normally this will be the goose parent, but he could imprint goslings on boots and they would then follow anyone wearing the boots. The instinct to respond to a parent was studied by Harry Harlow in the 1960s with a series of experiments so brutal they prompted legislation. He found that baby rhesus macaques became irreversibly disturbed if deprived of contact with a mother figure. They would seek affection in preference to food and in spite of negative stimuli such as pain – so instinct could be stronger than conditioning.

Lorenz's imprinted goslings treated him as though he were a parent.

1909

NITRATES FROM NOWHERE: THE HABER PROCESS

'It is a very strange thing to reflect that, but for the invention of Professor Haber, the Germans could not have continued the war after their original stack of nitrates was exhausted.'

– Winston Churchill, 1925

Fritz Haber in 1922.

Plants need nitrogen to grow, and adding nitrogen to the soil encourages growth. For centuries, this has been supplied in fertilizers such as manure, and through planting crops such as beans that help to fix nitrogen in the soil. At the end of the 19th century, the population was growing quickly, outstripping demand for conventional fertilizers – particularly in Germany, which has areas of infertile soil. At the time, the main source of fertilizer was from saltpetre (potassium nitrate), imported from Chile. This was not a secure source. The insecurity of the supply prompted German chemists, including Fritz Haber, to look for an alternative.

Nitrogen and hydrogen react together under pressure and at high temperatures to produce ammonia gas. The product and reactants are in equilibrium, the reaction going backwards and forwards.

By 1909, Haber could produce ammonia, but only drop by drop. He set out to find the best conditions to favour the ammonia side of the equilibrium. He concluded that mixing the gases at 450°C (842°F) and 250 atmospheres pressure in the presence of a catalyst (either osmium or uranium) produced the best yield of ammonia. In 1910, Carl Bosch started work on an industrial scale for fertilizers. It was the most important chemical process developed in the 20th century, and would go on to feed the world. Unfortunately, nitrates are also an ingredient of explosives and would be put to far deadlier use in the coming world wars.

FLASHPOINT FACT

Nitrogen makes up 78 per cent of the air we breathe, and hydrogen is the most abundant element in the universe, so there is no shortage of the gases needed to make ammonia.

TIMELINE

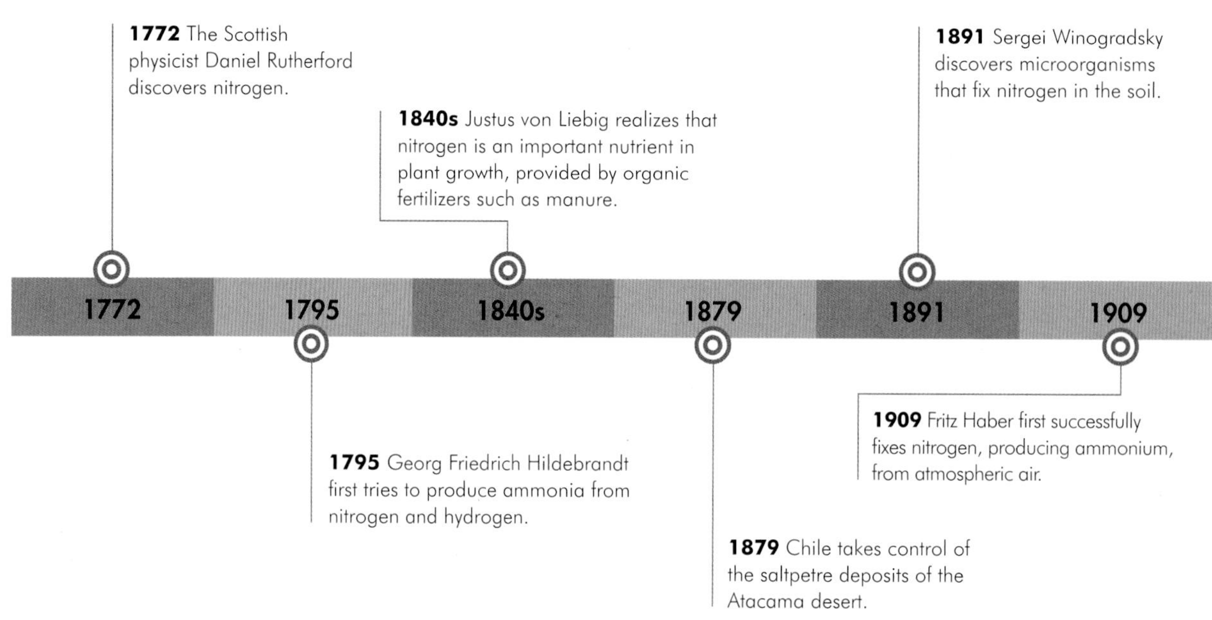

1879
THE NEED FOR NITROGEN

Following the 'Saltpetre War' of 1879 between Chile, Bolivia and Peru, Chile took charge of the Atacama desert and its rich supply of saltpetre, used to supply nitrogen fertilizer. As the supply was controlled by an oligopoly, the price was forever rising. By 1900, Chile produced two-thirds of the world's fertilizer. Saltpetre was crucial to European food security. The demand for saltpetre was particularly keen in Germany, where the soil was of poor quality and needed constant fertilizing. This led German chemists to lead the quest for economical ways of manufacturing ammonia. German consumption of saltpetre rose from 350,000 tonnes in 1900 to 900,000 tonnes in 1912, before industrial-scale production with the Haber process reduced dependency on saltpetre.

1914
THE WORLD AT WAR

Although spurred on by the need for fertilizer, the development of the Haber-Bosch process gave Germany a huge advantage in the First World War. Explosives such as nitroglycerine and TNT are nitrates, so British blockades stopped the import of saltpetre to reduce Germany's ability to make explosives. The industrial-scale use of the Haber process, though, meant that Germany had an almost unlimited supply of nitrates to use in weapons. It has been estimated that without the Haber-Bosch process, Germany would have run out of explosives at some point in 1916, bringing the war to a much earlier end and saving millions of lives.

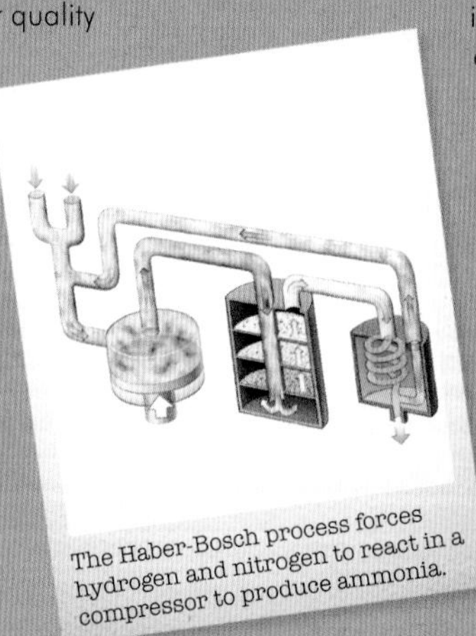
The Haber-Bosch process forces hydrogen and nitrogen to react in a compressor to produce ammonia.

1910 Haber sells his process to the firm BASF, which appoints Carl Bosch to develop it industrially.

1914 British naval blockades prevent Germany importing saltpetre from Chile; production of nitrates using the Haber-Bosch process becomes all-important.

1920 Haber and Bosch share the Nobel Prize for Chemistry for their work on the Haber-Bosch process. Bosch is not able to receive the prize at a public ceremony because his role in developing the chlorine gas used as a poison in the First World War has made him internationally despised.

1936 Controlled-nitrogen-release fertilizers are developed, able to trickle nitrogen into the soil at a steady rate over a period of time.

2000 Use of nitrogen fertilizers reaches around 100 billion kilograms a year, worldwide.

1910 1914 1920 1936 2000

1918
FEEDING THE WORLD

After the First World War, nitrate production was focused on use in fertilizers. The Haber-Bosch process then fuelled the explosion in agricultural production that sustained population growth throughout the 20th century. Ammonia was produced in vast industrial production plants, built all over the world. Even communist China, devastated by the famines caused by disastrous farming policies in the 1950s, began to build Haber-Bosch ammonia-production plants.

Ammonium production lies at the heart of the fertilizer industry.

2015
STILL GOING

Today, more than 40 per cent of the world's protein supply is produced by means of the Haber-Bosch process providing nitrate fertilizers. The conditions used in current industrial production are 450 degrees, 200 atmospheres and a catalyst of iron and potassium hydroxide. In these conditions, ammonia production is about 15 per cent; the ammonia is condensed by cooling and the unused gases recycled for another go, eventually producing 98 per cent conversion. A lower temperature would produce more ammonia per pass, but more slowly. The catalyst speeds up the reaction, too. The hydrogen is derived from natural gas (methane) and the nitrogen from the air.

1911

DISCOVERING THE STRUCTURE OF THE ATOM

The Cavendish laboratory at Cambridge University, where Rutherford worked on the structure of the atom.

'[It was] quite the most incredible event that has ever happened to me in my life. It was almost as incredible as if you fired a 15-inch shell at a piece of tissue paper and it came back and hit you.'

– Ernest Rutherford

Rutherford with his apparatus for working with alpha particles.

The now-familiar model of the atom that has a nucleus orbited by electrons originated with the New Zealand-born physicist Ernest Rutherford in 1911. During one of his many experiments on radioactivity, he directed Ernest Marsden and Hans Geiger firing a beam of alpha particles (helium nuclei) at a very thin sheet of gold foil. As they expected, most passed straight through – but a few bounced off. This result was stunning, and forced Rutherford to reconsider the structure of the atom as it was then conceived.

Instead of the atom being of fairly uniform construction, Rutherford proposed that the vast majority of its mass is concentrated into a very dense but tiny nucleus and most of the rest of it is empty space, with electrons orbiting the nucleus at a considerable distance. The few alpha particles that had bounced back had collided with and been repelled by a positively charged nucleus. He calculated that the nucleus represents only about about 1/10,000th of the radius of the atom. This was an astonishing discovery. It means that, if the nucleus were a beach ball 40cm (16in) wide, the outer electron would be 4km (2.5 miles) away.

Rutherford's nuclear model of atomic structure was revolutionary in more ways than one, as it introduced the idea that atoms have parts – that there are subatomic particles. He went on to isolate the proton in 1920 and propose the existence of the neutron, the other particle in the nucleus. The electron had already been discovered by J.J. Thomson in 1897.

FLASHPOINT FACT

If the nucleus of an average-sized atom were 60m (197ft) across, the whole atom would be the size of the Earth.

TIMELINE

1815 William Prout suggests that all atoms are made up of combined hydrogen atoms.

1894 G.J. Stoney proposes the name 'electron' for 'atoms of electricity'.

1897 J.J. Thomson discovers the electron.

1904 J.J. Thomson proposes the 'plum pudding' model of the atom.

1904 Hantaro Nagaoka describes the planetary model, with a large nucleus surrounded by electrons orbiting in rings in the same plane.

1905 Einstein's paper on Brownian motion suggests a proof that atoms and molecules exist.

FLASHPOINTS

1904
THE ATOMIC PLUM PUDDING

Thomson had discovered the electron in 1897, the first subatomic particle to be recognized. In 1904, he proposed a model of the atom in which a cloud or soup of positive charge is studded with charged 'corpuscles' carrying a negative electrical charge. Others described this as a positively charged 'pudding' dotted with negatively charged 'raisins'. The same year, the Japanese physicist Hantaro Nagaoka proposed a planetary model of atomic structure, with the atom surrounded by electrons orbiting in the manner of Saturn's rings. His predictions of a comparatively large nucleus and orbiting electrons were both correct, but the detail was wrong.

1913
SHELLS WITHIN SHELLS

The Danish physicist Niels Bohr extended Rutherford's model of the atom in 1913, in line with quantum theory. In his model – which is still basically current – electrons orbit the nucleus in clearly defined 'shells' or spaced orbits. The energy associated with particular orbits varies, but only in line with fixed quanta (parcels) of energy. If an electron jumps from one orbit to one with a lower energy, a quantum of energy is released. It is for this reason that atoms emit light in fixed wavelengths: a photon with a certain amount of energy is emitted as an electron moves between shells. Their being contained in shells also explains why they do not immediately fall towards the positive nucleus, a puzzle that Rutherford could not solve.

J.J.Thomson's model proposed the atom as something like a pudding studded with raisins.

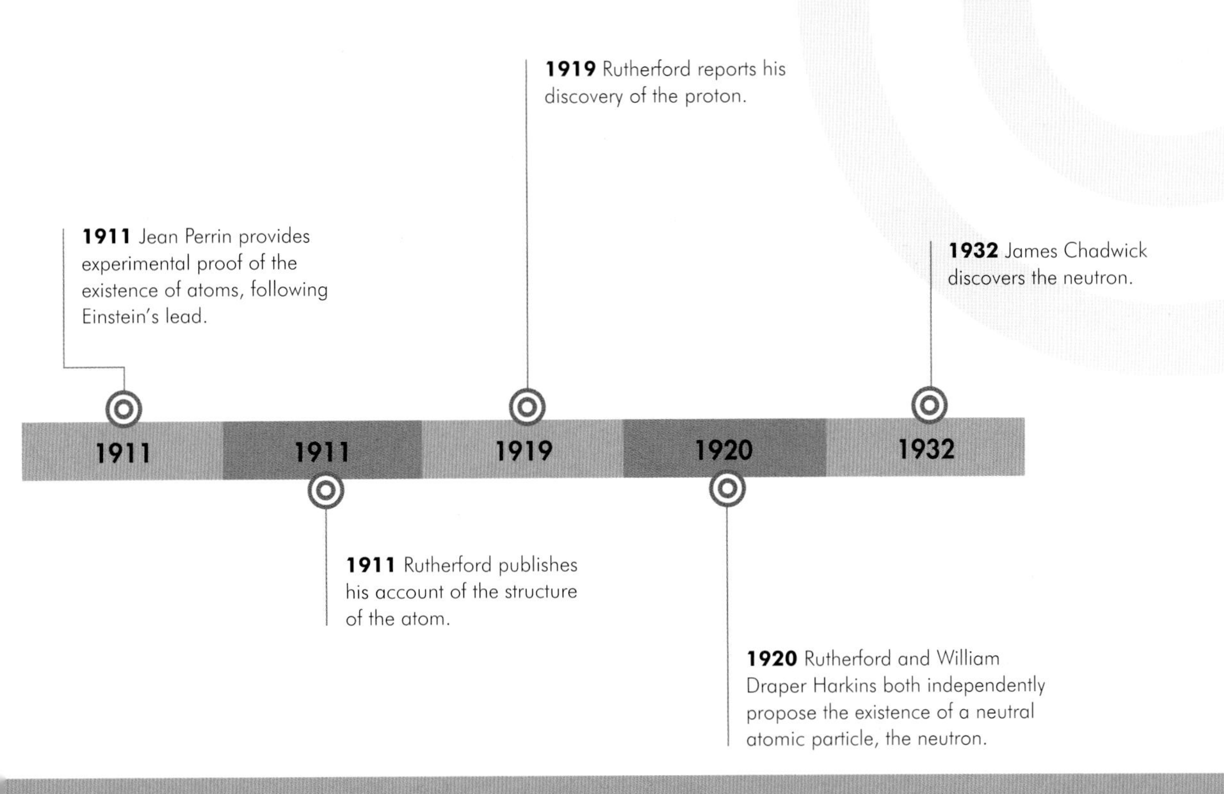

1924
LIGHT FROM MATTER

Although Bohr worked out that electrons release or absorb a quantum of energy when jumping between shells, he offered no explanation as to why this happens. That piece of the puzzle was found by Louis de Broglie in 1924. De Broglie suggested that any particle had an associated wavelength, calculated by dividing Planck's constant by the momentum of the particle. When an electron drops to a lower atomic orbit (closer to the nucleus), its frequency changes, and the extra energy is disposed of by emitting a photon. When Louis de Broglie won the 1929 Nobel Prize in Physics for his work, it was the first time the prize had been awarded for a PhD thesis.

An atom emits a photon as an energized electron drops a level in the orbital shells.

1932
MORE SUBATOMIC PARTICLES

After discovering the atomic nucleus in 1911, Rutherford assumed that it was made of a positively charged particle which balanced the negative charge on the electron. In 1917, he demonstrated that firing alpha particles (helium nuclei) at nitrogen produced hydrogen nuclei. (This was the first nuclear reaction.) Neither experiment nor theory were consistent with the nucleus containing only a single type of particle, though, so in 1920, he suggested that the nucleus consisted of positively charged protons and some type of neutral particle. It was James Chadwick, examining a new type of radiation, who discovered the neutron in 1932 – an uncharged particle of about the same mass as the proton.

A cloud-chamber photo taken in 1932 shows the track of a proton recoiling after being struck by a neutron.

1912

WEGENER AND THE STRUCTURE OF THE EARTH

'Scientists still do not appear to understand sufficiently that all earth sciences must contribute evidence toward unveiling the state of our planet in earlier times.'

– Alfred Wegener, *The Origins of Continents and Oceans*, 1915

Alfred Wegener's diagrams, published in 1922, show how he believed the continents had moved through geological history.

Geophysicist Alfred Wegener was struck by the strange fact that fossils of similar plants and animals are found in South America and in Africa. Further, the east coast of South America looks as though it would fit snugly against the west coast of Africa. From these clues, Wegener developed his theory of continental drift, which states that the landmasses move slowly around the Earth. Africa and South America were, he claimed, once adjoining, and have drifted apart over a period of millions of years. Further evidence for his theory came from the similar geology of South America and Africa.

His theory met a hostile reception when he announced it in 1912 and published it in 1915. He proposed that all the land in the world had once been clumped into a single super-continent, which he named Pangaea, and that this began to break up around 200 million years ago with the chunks of land drifting slowly towards their current positions. Not only did his ideas clash with the Christian belief that the world was unchanging and had been since Creation, but he could propose no mechanism for movement of the landmasses. In fact, later work on the structure of the Earth would provide the mechanism. The continental plates holding the land and oceans are floating on a thick layer of very hot, semi-liquid magma, the Earth's mantle, which slowly moves around the planet. His theory was largely derided until the 1960s, but then proven conclusively when continental drift was measured in 1984.

FLASHPOINT FACT

Before continental drift and tectonics explained how mountain ranges form, people believed they were wrinkles that appeared as the cooling Earth shrank.

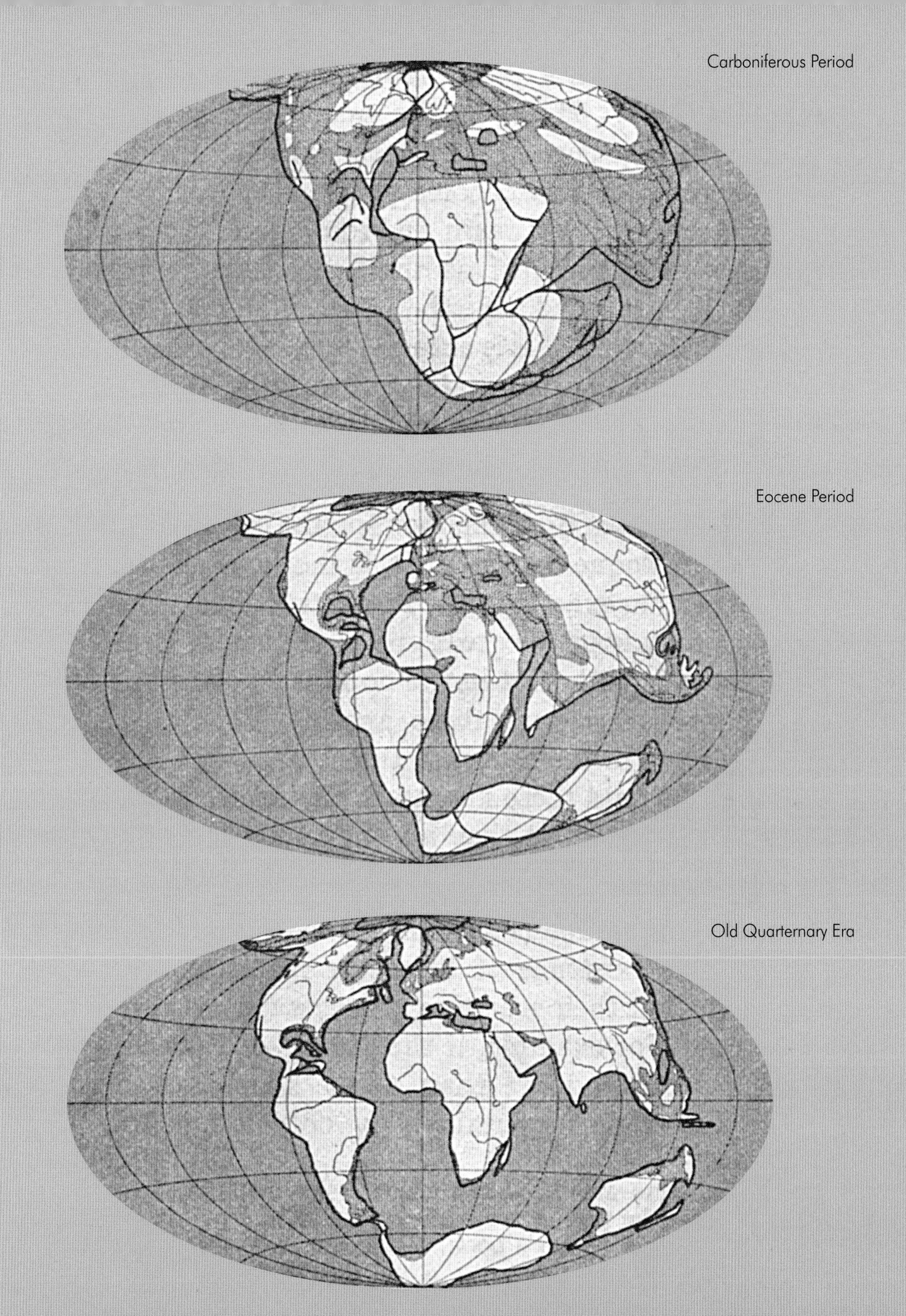
Carboniferous Period
Eocene Period
Old Quarternary Era

TIMELINE

1654 Archbishop Ussher declares that the Earth was created in 4004BC and had remained unchanged since; his calculation was based on genealogies in the Old Testament.

1880 The seismograph is invented to measure the seismic waves generated by earthquakes.

1907 Bertram Boltwood publishes his findings working with radioactive rock, the oldest of which he found to be 570 million years old.

1912 Alfred Wegener outlines is theory of continental drift, but it was not popular.

1924 Arthur Holmes devises radiometric dating (using the radioactive decay of uranium) and calculates the age of Earth as at least 1.6 billion years.

1935 Charles Richter publishes a logarithmic scale to measure the amplitude of the seismic waves, which is used to monitor seismic activity, predict earthquakes and report their intensity.

1654 | 1880 | 1907 | 1912 | 1924 | 1935

FLASHPOINTS

1936
INSIDE THE EARTH

Until 1936, scientists believed that the Earth had a liquid core and a solid mantle and crust. But then Inge Lehmann overturned this with her measurements of seismic waves (waves that travel through the Earth after a jolt such as an earthquake or explosion). By analyzing the type of waves recorded by seismographs in Europe at the time of an earthquake that had occurred in New Zealand in 1929, she was able to demonstrate that the Earth must have a solid inner core, surrounded by a liquid outer core.

1956
AGE-OLD EARTH

In 1654, Archbishop Ussher calculated the date of Creation to be 4004BC, a date he arrived at by adding together the ages given for figures in the Old Testament. In 1862, William Thomson published calculations putting the age of the Earth between 20 and 40 million years old, based on how long he thought it would take a blob of hot, molten rock to cool to the Earth's current state. There was considerable disagreement with other scientists who believed it to be a good deal older, but still nowhere near the current figure, calculated in 1956, of 4.55 billion years old. At the time of Wegener's work, the largest estimate for the age of the Earth was around 570 million years.

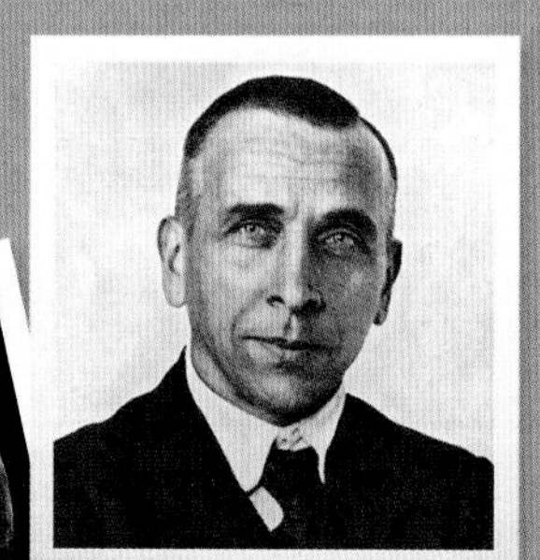

Alfred Wegener.

The interior structure of the Earth, showing the mantle and core.

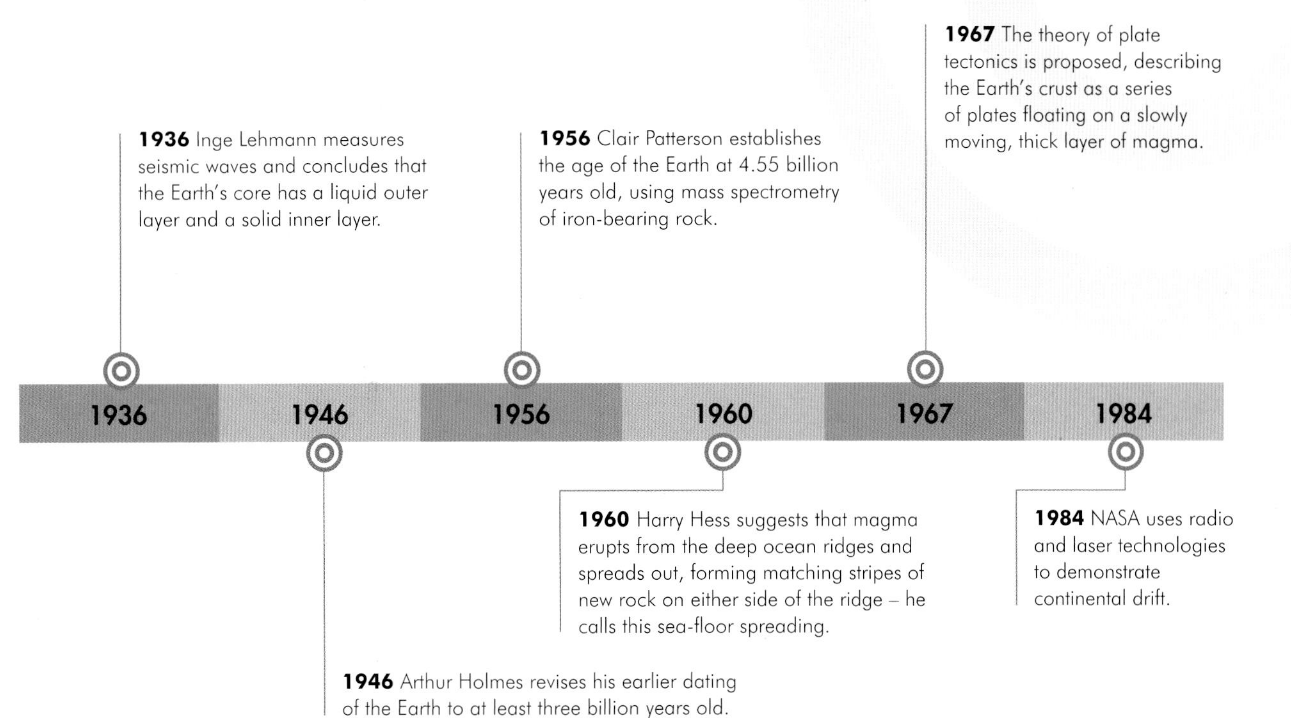

1960
THE SEA IS GROWING!

The discovery of sea-floor spreading in 1960 led to renewed interest in Wegener's theory of continental drift. The geologist Harry Hess had discovered hundreds of flat-topped, deep-sea mountains in the Pacific Ocean in 1946. When a great undersea mountain range was found bordering a huge canyon, the Great Global Rift, he came up with a theory to account for these and Wegener's continental drift. He suggested that hot magma oozed from the rift, hardening into new rock and forcing the continents apart. In 1962, he formed this into the theory of tectonic plates: the land and oceans are on plates of Earth's crust, floating on top of a thick layer of slow-moving magma.

1984
TECTONIC THEORY

Tectonic plate theory married Wegener's continental drift with other geological phenomena. Earthquakes occur when the plates grind against each other and jolt in their movement. Volcanic eruptions are fed by rock melting as one plate is forced under another, or from hotspots below the crust. And mountain ranges grow as plates collide and push against each other head-on, forcing rock upwards. The movement of the tectonic plates was finally confirmed in 1984 by NASA measuring the movement: Europe and America are moving apart at a rate of 1.5cm (0.6in) per year and Australia is moving towards India at 7cm (2.8in) per year.

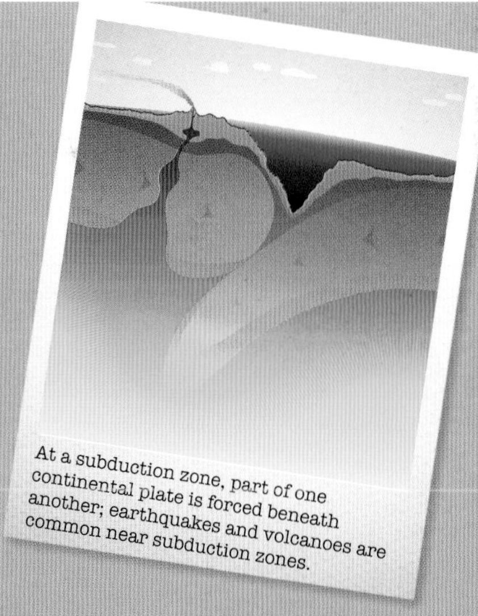
At a subduction zone, part of one continental plate is forced beneath another; earthquakes and volcanoes are common near subduction zones.

1915

EINSTEIN AND THE THEORY OF RELATIVITY

'It has become appallingly obvious that our technology has exceeded our humanity.'

– Albert Einstein

Albert Einstein lecturing at the Carnegie Institute of Technology in 1934.

Albert Einstein (1879–1955) must be the most famous scientist of the 20th century, though very few people understood his ideas when he first stated them.

Einstein felt that Newtonian physics had limitations when applied to new developments, and set out to correct this. His great breakthrough came in 1905, his *annus mirabilis* ('miraculous year'). He published four groundbreaking papers – on the photoelectric effect, the movement of atoms, special relativity and the equivalence of energy and mass. Together, these moved physics into a new era.

His theory of special relativity demonstrated that the speed of light is absolute – it is the same regardless of the position of the observer – but that time and space are relative, changing with the conditions of the observer. Travelling at great speed (near the speed of light) makes time pass more slowly.

Einstein was aware that the theory did not take account of gravity. He spent the next ten years wrestling with the very complex maths he needed to make it generally applicable throughout the universe. Finally, in 1916, he published the theory of general relativity. In it, he explains gravity as a distortion of the space–time continuum caused by the presence of massive objects. A massive star bends space and time, rather like a heavy ball makes a dip in a blanket pulled taut. The effect is that other bodies of lesser mass 'fall' towards the heavier mass.

FLASHPOINT FACT

After his death in 1955, Einstein's brain was kept in a jar until 1998. The pathologist at Princeton Hospital, Thomas Harvey, had removed and kept it without permission.

Impuls
Kin. Energ.
$x = \frac{x' + vt'}{\sqrt{1-v^2}} \quad y = y' \quad z = z'$
Hyp.

TIMELINE

1827 Robert Brown reports the random movements of pollen grains in water, now known as Brownian motion.

14 March 1879 Albert Einstein is born in Ulm Württemberg, Germany.

1902 Unable to find any work as a teacher or academic, Albert Einstein takes a job as a clerk at the Swiss Patent Office.

1905 In this *annus mirabilis* – miraculous year – Albert Einstein publishes four major theoretical papers, including the first published exploration of the Theory of Special Relativity and the first formulation of the famous equation $E=mc^2$.

1915 Einstein completes his General Theory of Relativity.

29 May 1919 A solar eclipse provides dramatic observable evidence that Einstein's General Theory of Relativity is correct.

1827 | 1879 | 1902 | 1905 | 1915 | 1919

FLASHPOINTS

1879–1955
A LIFE IN BRIEF

Einstein was no child prodigy. He was slow to talk and failed the entrance exam for university. After graduating from the polytechnic in Zurich, Switzerland, he could not find teaching work and took a job in the Patent Office. He pursued his interest in physics in the evenings, and gained his PhD in 1905.

The General Theory of Relativity established his global reputation. He settled in the USA in 1935 to escape persecution by the Nazis, becoming an American citizen in 1940. Despite being a pacifist, he worked on the Manhattan Project to develop nuclear weapons, as he feared Germany would develop the bomb first. He spent his last years seeking an elusive unifying theory that would bind all physics together.

1905
ATOMS, MATTER AND ENERGY

Einstein's paper on the movement of atoms explained Brownian motion – the movement of pollen particles in water observed by Robert Brown in 1827. Einstein explained that the pollen is jostled by molecules of water. This was definitive proof of the existence of atoms and molecules, previously proposed but not demonstrated.

Einstein's final paper of 1905 was on the equivalence of energy and mass. It contains the first version of his famous equation, $E=mc^2$. The fact that matter can be converted to energy makes nuclear power and nuclear weapons possible.

Solar panels make use of the photoelectric effect to generate electricity from sunlight.

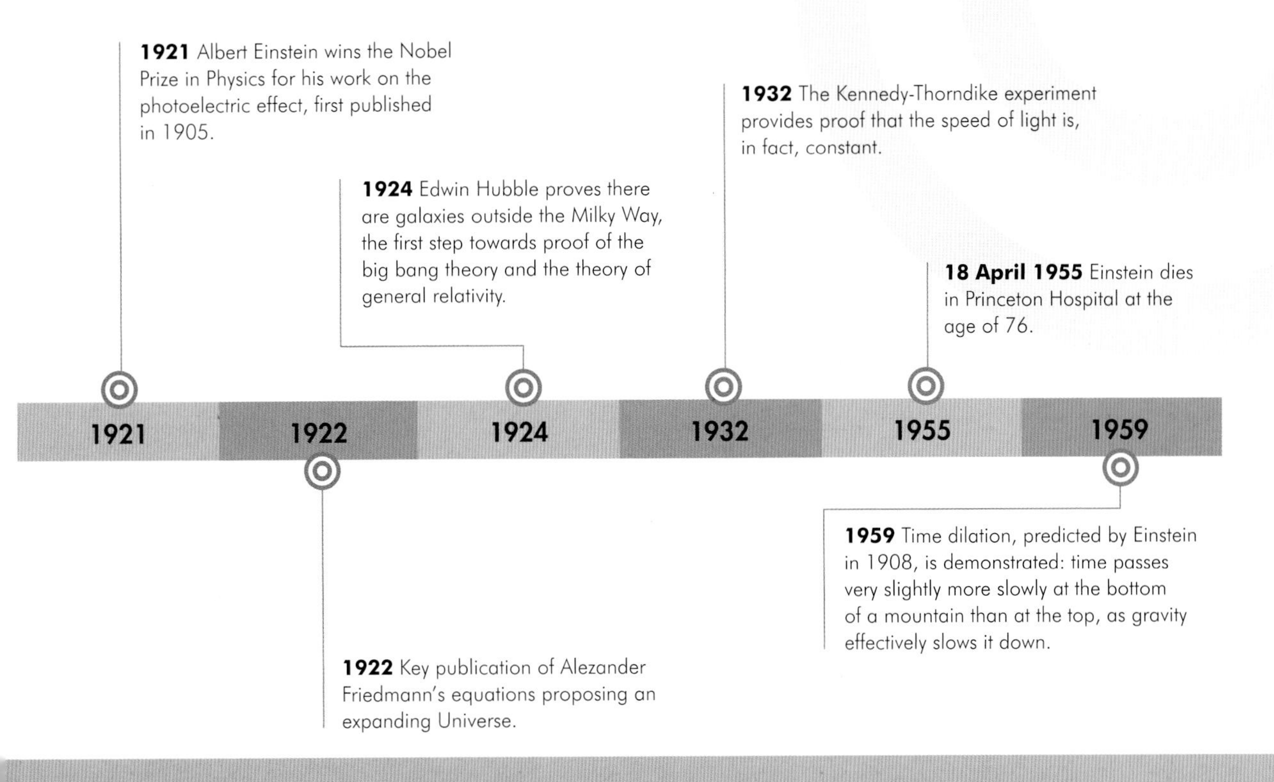

1919
PARCELS OF LIGHT AND ELECTRICITY

The four papers of his *annus mirabilis* made Einstein's name. The first, on the photoelectric effect, used Max Planck's theory of quanta, or discrete packets of energy that cannot be divided. Einstein's explanation of the photoelectric effect relied on light quanta-displacing electrons (which we can think of as electricity quanta) and so producing an electric current. The idea that light is formed of quanta met considerable resistance until Robert Millikan proved it in 1919. The photoelectric effect is exploited in solar power cells, which convert the light energy of the sun into electricity by this means.

A photograph of a total solar eclipse, taken by Edington on 29 May 1919.

1919
PROOF OF GENERAL RELATIVITY

The proof of Einstein's General Theory of Relativity came in 1919. Einstein had proposed in 1911 that during an eclipse, a star behind the sun should be visible, its light skirting around the sun, bent by the sun's gravity. Normally, the sun is too bright for the effect to be visible, but when it is darkened by a total eclipse that obstacle is removed. Arthur Eddington, Cambridge professor of astronomy, sailed to Principe off the coast of Africa to photograph the eclipse. Despite cloudy weather for the first 400 seconds, he took a conclusive photograph in the last ten seconds: Einstein's account of gravity is closer to reality than Newton's.

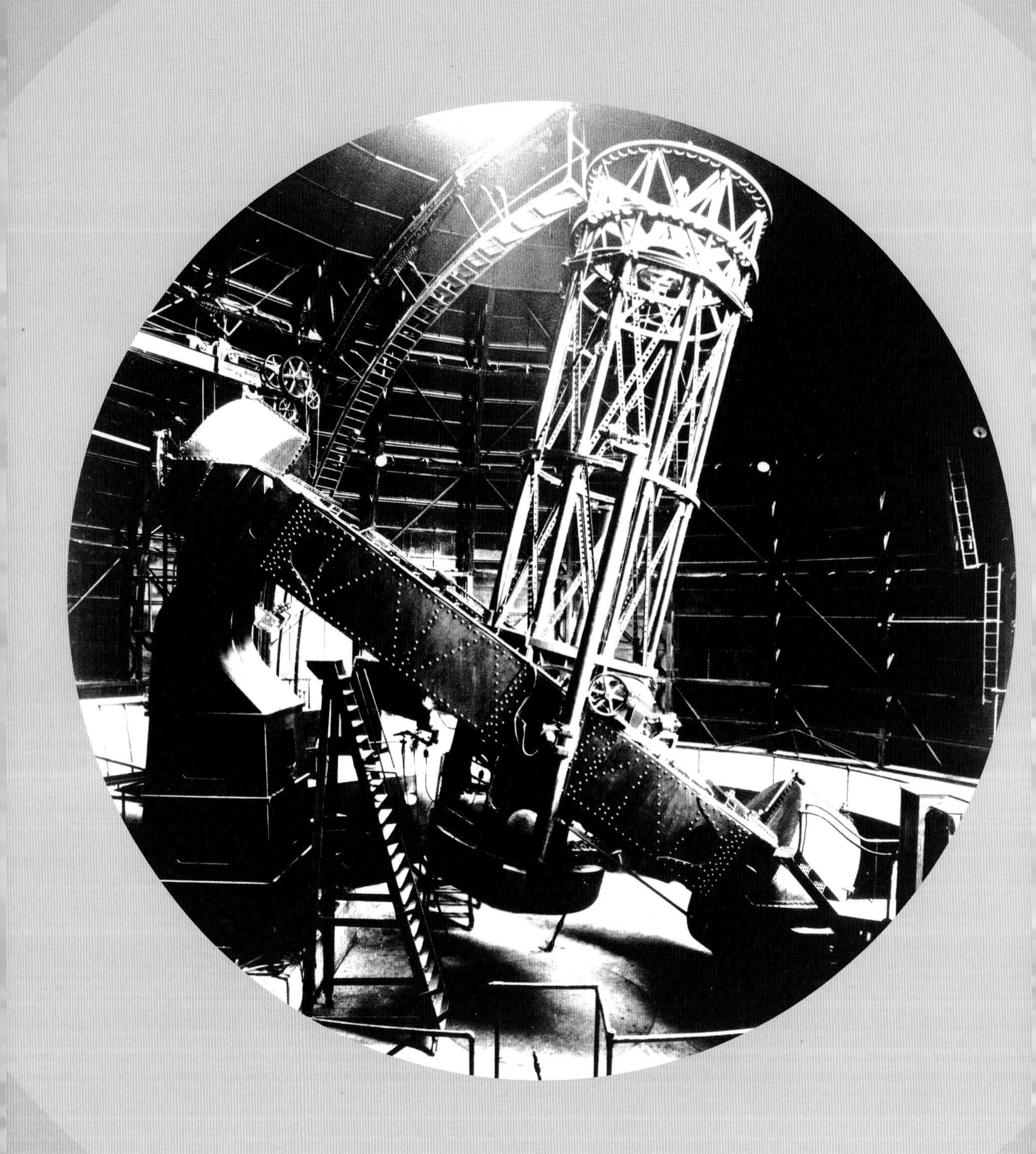

1920–1939

The years between the First and Second World Wars were turbulent and chaotic. The dominance of German science seen in the first 15 years of the century diminished, with the country in economic and political turmoil. The great German physicists either fled or were commandeered by the rising Nazi machine. The greatest of them would end up in the USA, either before or after the Second World War.

The 1920s saw humankind looking to the stars with new interest. Building on Einstein's work of the previous decades, physics began to extend its reach into space, with the start of current theories about the origins of the universe and the first radio telescopes. These decades were dominated perhaps for the ordinary person by advances in communications brought about by harnessing the electromagnetic spectrum – the telephone and radio became household objects and the television first appeared.

Other developments that would prove a springboard for great revolutions later in the century included the first work on computers, the first plastics and the discovery of the first antibiotics. Unlike the discoveries of the first 20 years of the century, those of the 1920s and 1930s would have a massive impact on the lives of all ordinary people in the decades to come.

1920

PUTTING RADIO TO WORK

'For God's sake, go down to reception and get rid of a lunatic who's down there. He says he's got a machine for seeing by wireless! Watch him – he may have a razor on him.'

– News editor of *The Daily Mail* on John Logie Baird, visiting with his prototype television in 1925

Guglielmo Marconi and his assistant George S. Kemp were the first to transmit and receive radio signals over the Atlantic Ocean.

The electromagnetic spectrum is energy of different wavelengths, ranging from radio waves through visible light in the middle to X-rays. Radio waves have the longest wavelength and X-rays the shortest. While X-rays were the first to be put to work, with medical X-rays beginning in 1896, radio has been the most exploited.

The first use of radio was to transmit simple signals in Morse code. With the discovery in 1906 that it could be used to transmit the human voice, the whole area of modern communications opened up. The first regular public voice broadcasts began in Argentina in August 1920, quickly followed by KDKA in Pennsylvania, USA, in October. It rapidly took off, with commercial radio sets soon common in homes. 'Ham' radio enthusiasts, who built their own sets and transmitters, continued and flourished alongside the state and commercial broadcasters. When John Logie Baird discovered a way of encoding images as well as sound, television became the next use for radio waves. Since then, radio has been put to work with intercoms, mobile phones, CCTV and the wireless internet.

Radio, television and data transmissions work by encoding images and other types of data into radio waves, changing the frequency and wavelength to carry information. The data is encoded by a transmitter and decoded at the other end by a receiver, which separates the carrier wave from the information and rebuilds the information into its original format.

Like all other types of electromagnetic radiation, radio waves travel at the speed of light. Although we use radio for sound, the transmission is much faster than the speed of sound because it is carried in the same form as light. Unlike sound waves, radio waves can travel through space, making communication with spacecraft possible and allowing waves to be 'bounced' off satellites, giving us satellite TV and navigation systems.

FLASHPOINT FACT

On the farms of the American west, barbed-wire fences were sometimes used as radio antennae – sometimes accidentally, with farmers hearing broadcasts as they walked their fields.

TIMELINE

1844 First long-distance electric telegraph line opens between Washington, DC and Baltimore.

1873 James Clerk Maxwell publishes his theory of electromagnetism.

1888 Heinrich Hertz produces electromagnetic waves in the laboratory, but doesn't expect the process to be very useful.

1894 Sir Oliver Lodge sends the first message using radio in Oxford, England.

1895 Wilhelm Röntgen discovers X-rays by accident.

1901 Guglielmo Marconi transmits the first transatlantic radio signal.

1904 The first demonstration of radar, in the form of Christian Hülsmeyer's 'Telemobiloscope'.

1906 Reginald Fessenden, first uses radio to transmit sound – live music and readings broadcast to ships in the North Atlantic.

FLASHPOINTS

1672
LIGHT AND DARK

The first part of the electromagnetic spectrum to be investigated was, not surprisingly, visible light. Isaac Newton started the ball rolling with his discovery that white light is a composite of many colours of light, demonstrated in 1672. Infrared and ultra violet were discovered early in the 19th century. But it was not until James Clerk Maxwell worked out the nature of electromagnetic radiation in 1873 that the link between phenomena as apparently diverse as electricity, magnetism, visible light and radiant heat began to emerge. Maxwell proposed that electromagnetic disturbances would propagate through space at the speed of light in the form of waves. Heinrich Hertz confirmed this experimentally in 1887–8.

1894
CANNY PHOTONS

The earliest form of telegraphy sent electrical signals along wires to transmit Morse code, a simple binary code of dots and dashes. This required wires to be in place wherever messages were to be transmitted, so was soon superseded when 'wireless telegraphy', using radio, came along in the 1890s. When Guglielmo Marconi successfully transmitted a signal across the Atlantic in 1901, the age of radio had begun. It took only five years to develop the first system to transmit sound as radio, achieved by coding it into varying amplitude (height) of the radio waves. Radio became popular with 'ham' radio enthusiasts.

A Marconi radio tuner of the type used on the ill-fated *Titanic*.

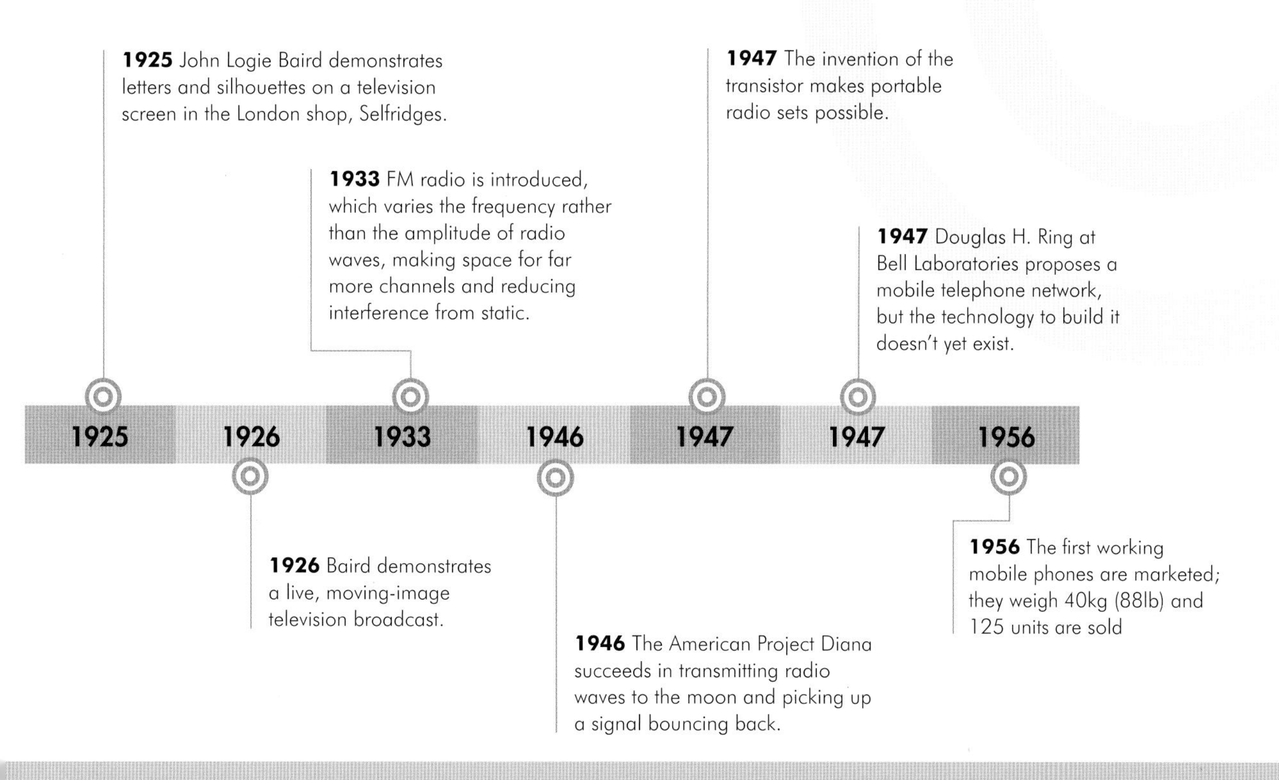

1939
SOUND AND VISION

From the 1920s, radio and television began to shape communications and media. They were used not only for public information and entertainment, but also for military and scientific applications. One of the most important military applications is radar.

Radar works by detecting the reflection of radio waves beamed from a transmitter, and bouncing off a metal object such as an aircraft or ship. Initially developed in Germany in 1904, radar did not fully take off until 1939. The British pioneer of radar, Robert Watson-Watt, worked on its development from 1935; he is credited with playing a major role in the success of the Allied forces in the Second World War.

1950s
OUT OF THE BOX

For most of the population, the importance of radio waves from the 1950s onwards was in providing the immensely popular media of radio and television, which revolutionized leisure and communications. They gave access to current affairs and entertainment, and offered advertisers the opportunity to reach millions simultaneously. Other vital developments with radio waves were underway at the same time. The idea of mobile telephones, first proposed in the 1950s, became a tangible reality for most people in the 1980s and 1990s. As communications satellites filled the skies, the electromagnetic spectrum gave us not just mobile phones, but cable and satellite TV, GPS systems, wireless computer networks and the wireless internet. It is at the heart of the modern world.

Guglielmo Marconi.

1928
ANTIBIOTICS

'I have been trying to point out that in our lives chance may have an astonishing influence and, if I may offer advice to the young laboratory worker, it would be this—never neglect an extraordinary appearance or happening.'

– Sir Alexander Fleming, on his accidental discovery

Fleming's drawing of the plate on which he first found clearing *Staphylococcus* bacteria.

Antibiotics are one of the most important developments of the 20th century. They have saved millions of lives and yet their discovery was accidental.

The first modern antibiotic to be discovered and produced in large quantities was penicillin. The story of its accidental discovery by Alexander Fleming is well known. In 1928, Fleming took a holiday from work in his laboratory, piling up used petri dishes before he went, but not cleaning them out (he was notoriously untidy). His poor housekeeping was serendipitous. When he returned a couple of weeks later, his discarded dishes were still there. But there were clear spaces where the colony of the *Staphylococcus* bacteria he was growing had died. Investigation revealed that a mould had grown on the plates – *Penicillium notatum* – and this had killed the bacteria. Fleming found that the mould would kill a number of bacteria, including those causing diphtheria, scarlet fever, pneumonia and meningitis. He couldn't produce penicillin in large enough quantities to use as medicine, and he believed that its action was so slow that it would not remain in the body long enough to have an effect.

Ernst Chain and Howard Florey took up where Fleming had left off in 1938. They made a medicine from penicillin which they first tried in 1940 on a patient who had developed an infection after being injured gardening and was close to death. He began to improve, but after five days the supply of penicillin ran out and he died.

In 1941, Chain and Florey were working in the USA when a laboratory assistant bought a melon infested with *Penicillium chrysogeum*. It turned out to produce 200 times as much penicillin as *Penicillium notatum*. After processing, it became the source of large enough supplies of penicillin to save many Allied lives in the Second World War.

FLASHPOINT FACTS

Alexander Fleming amused his colleagues by creating pictures on petri dishes, growing different-coloured moulds to make up different parts of the design.

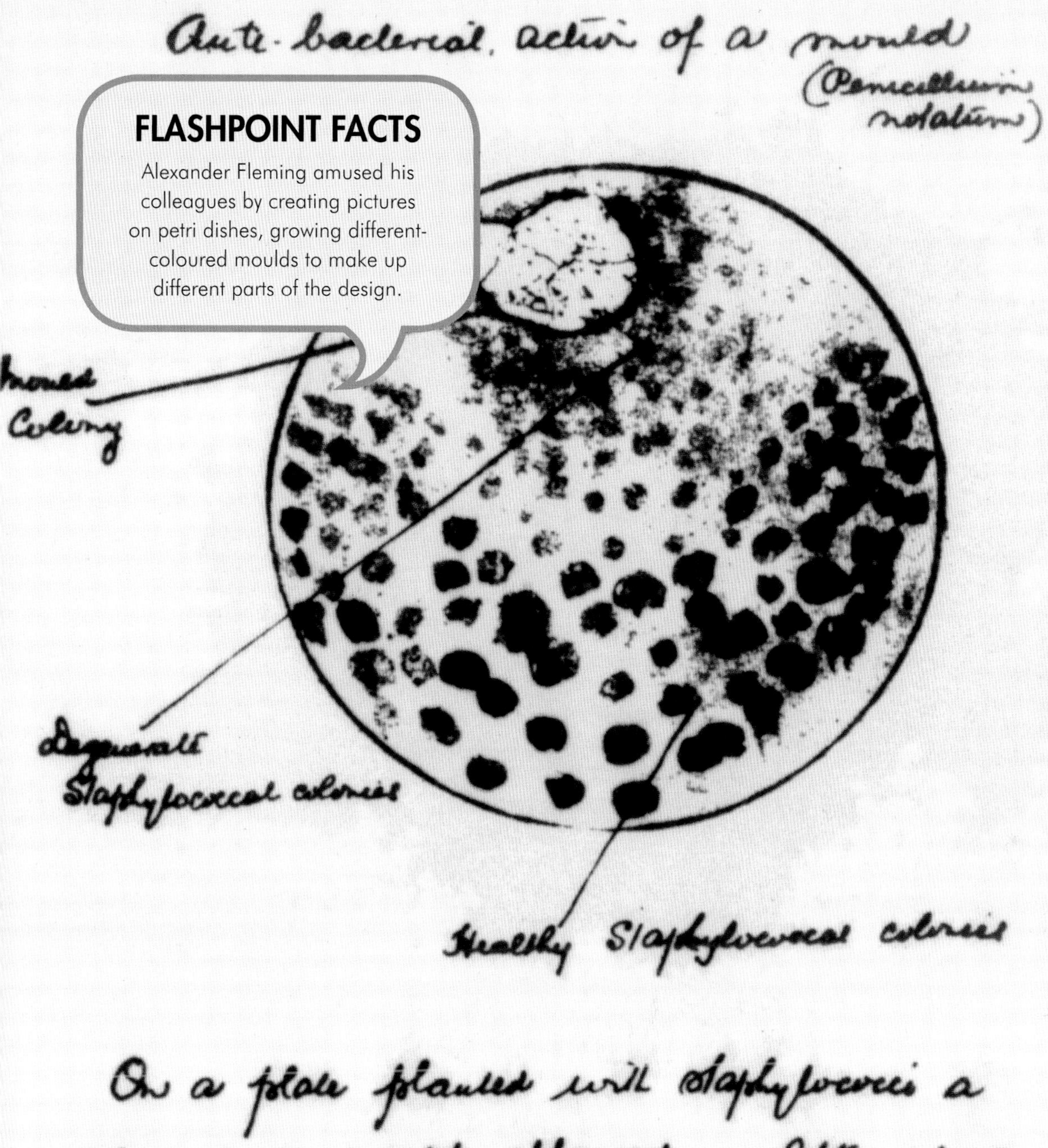

On a plate planted with staphylococci a colony of a mould appeared. After about two weeks it was seen that the colonies of staphylococci near the mould colony were degenerate

TIMELINE

1896 Ernest Duchesne discovers penicillin while a final-year student at the Pasteur Institute in Paris, but it goes unnoticed.

1860–4 Louis Pasteur demonstrates that bacteria can cause disease – the first formal evidence of germ theory.

1909 Paul Ehrlich and Sahachiro Hata develop Salvarsan to treat syphilis. It is the first treatment targeted at a specific organism, calculated to affect it using detailed scientific knowledge – a 'magic bullet' for a specific condition.

1915 Frederick Twort discovers phages – viruses that attack bacteria.

1923 The Eliava Institute is founded in Georgia, USSR, to study the use of phages as a means of controlling bacterial disease.

1928 Alexander Fleming accidentally discovers the antibacterial effect of the mould *Penicillium notatum* on *Staphylococcus aureus*.

1935 Gerhard Domagk uses the dye prontosil rubrum on his desperately ill young daughter, saving her life. This was the first sulpha drug.

1896 | 1860–4 | 1909 | 1915 | 1923 | 1928 | 1935

FLASHPOINTS

Pre-1945
UNLUCKY ACCIDENTS

Before the development of antibiotics, even seemingly minor injuries and infections could prove serious or even fatal. If the body's immune system was not sufficiently robust to defeat invading bacteria, the outlook was grim. From the earliest times, people had used honey and oil on wounds – excluding air prevented infection by aerobic bacteria (those that need oxygen). Packing wounds with moss, initially probably intended as a dressing, also sometimes prevented infection, as the moss contained chemicals with antibiotic properties – but no one knew they were there, so it was rather hit and miss. The search for antibiotics began in earnest once the development of anaesthetics extended the potential of surgery. Once the pain was removed, infection was the largest barrier.

1935
PRONTOSIL TO THE RESCUE

The German chemist Gerhard Domagk had been experimenting with the dye prontosil rubrum when, in 1935, his six-year-old daughter pricked her finger on a knitting needle and developed a serious infection. The recommended treatment was amputation of the arm to try to prevent her death. Domagk knew that prontosil had antibiotic effects in mice, but it had never been tried with humans before. Desperate, he treated his daughter with it and she recovered – the first patient to receive an antibacterial agent. Prontosil and other sulpha drugs became immensely popular and saved many lives during the Second World War, but their frequent side effects meant that they were quickly replaced by later antibiotics.

Gerhard Domagk, 1949.

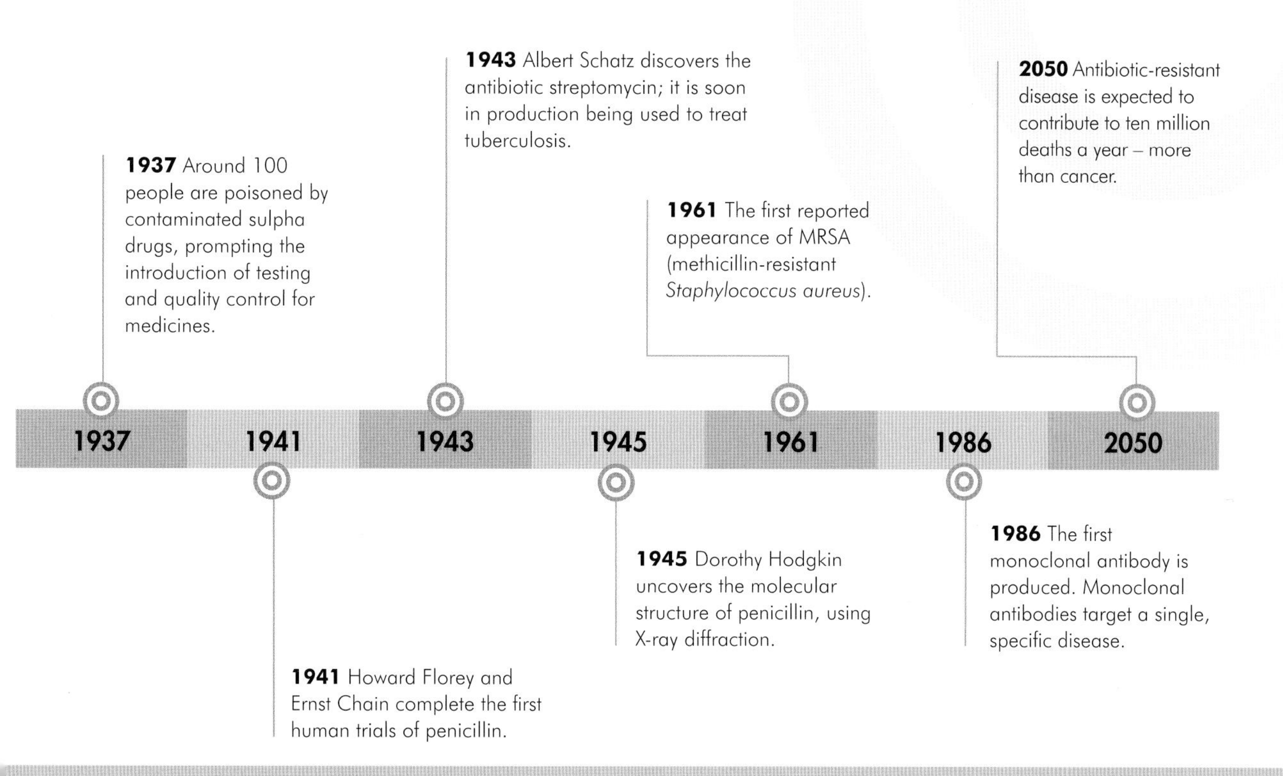

1945

PENICILLIN – FROM PLATE TO PRODUCTION

Although Domagk's daughter suffered her infection after Fleming's discovery of penicillin, it had not yet been developed into a medicine. That was achieved by Howard Florey and Ernst Chain in 1945. They, too, found that it was impossible to produce penicillin in large enough quantities from *Penicillium notatum*. They had succeeded in 1940 in delaying the death of their first patient, but then ran out of penicillin and he died anyway. After a worldwide search for a better mould for extracting penicillin, they finally found a source that yielded a thousand times that of Fleming's mould. Production started in vast fermentation towers in the USA, as English factories had been commandeered for the war effort. By 1945, 650 billion units were being produced each month.

1945

VARIATIONS ON A THEME

Bacteria adapt and evolve quickly, and it was not long before many bacteria developed resistance to penicillin: the wonder-drug no longer worked as well as it once had. The barrier to making variants of penicillin was understanding the structure of the molecule involved. The new technique of X-ray crystallography offered a chance. The complex task of breaking down the molecule, building an image of its components using X-ray crystallography, and then piecing them back together was finally accomplished by Dorothy Hodgkin in 1945. Armed with the knowledge of penicillin's molecular structure, chemists were able to manufacture effective artificial forms of penicillin and make new antibiotics.

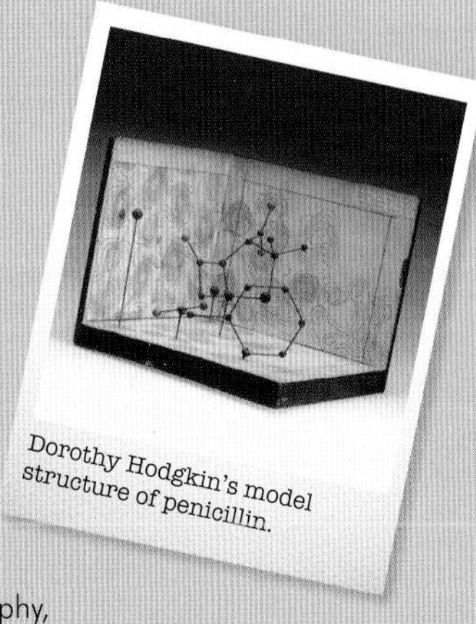

Dorothy Hodgkin's model structure of penicillin.

1928
SYNTHETICS BREAK OUT OF THE LAB

'As strong as steel, as fine as a spider's web.'

– DuPont advert for nylon

Wallace Carothers works in his laboratory at DuPont, *c.* 1927.

Although naturally occurring synthetics such as cotton and silk had been in use for centuries, the 20th century saw the development of artificial synthetics in many forms. The first, Bakelite, is a rigid and brittle material formerly used for items such as doorknobs, radio sets, gun parts and billiard balls. It fell out of favour in the 1940s when more versatile and flexible plastics appeared.

The first significant flexible artificial synthetic was produced by Wallace Carothers, who was working for DuPont in the USA in 1928. He was head of a team looking at artificial materials and had begun by investigating acetylene, a hydrocarbon (a chemical made principally from hydrogen and carbon). With Elmer Boton, he developed the artificial rubber neoprene, which went on sale in 1931. Carothers next set out to find a replacement for silk. Silk was essential for parachutes, ropes, tents and other items needed by the military; as US relations with Japan were breaking down, artificial silk became a top priority. That replacement was to be nylon, made from oil. It was extremely strong, yet light and flexible and could be spun into supple, thin fibres. One of the first commercial products was nylon stockings; 64 million pairs were sold in the first year of production.

Synthetics became the wonder materials of the 20th century. Nylon was followed Teflon, polythene, polystyrene and others.

The new synthetics proved incredibly versatile, being formed into everything from super-strong fibres to polyurethane paint and puffy extruded polystyrene good for packing materials and insulation. Whole new industries grew up around synthetic production and exploitation. In recent years, worries about the environmental impact of synthetics and their production from oil have led to the development of new polymers from plant material.

FLASHPOINT FACT

The polymer Teflon has a molecular mass of around 30 million – one of the largest molecules known.

TIMELINE

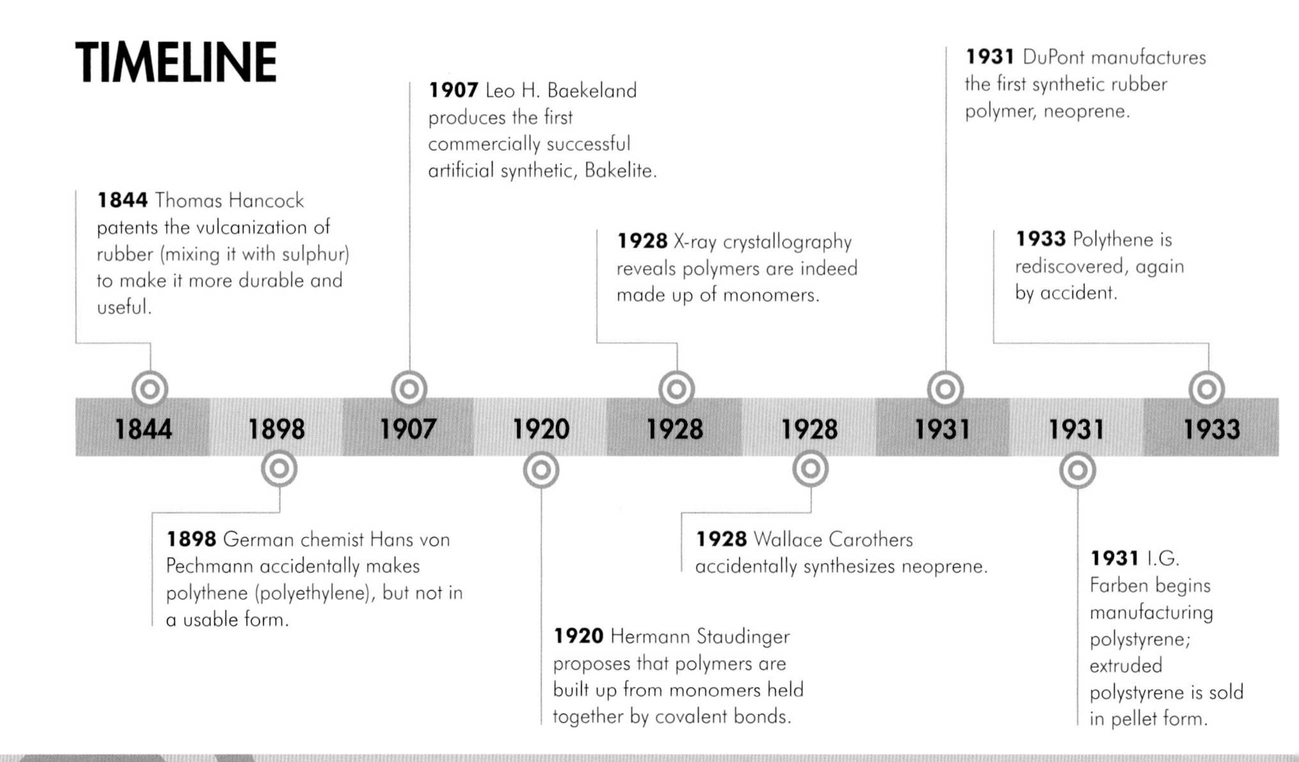

FLASHPOINTS

1920
MOLECULES LARGE AND SMALL

Hermann Staudinger was working on the synthesis of isoprene for the German chemical company BASF when he suspected it was the basis of the natural polymer rubber. He proposed that polymers are very large molecules (macromolecules), made up from smaller sub-units, called monomers. The monomers, he said, are held together by covalent bonds – a form of chemical bond formed by atoms sharing electrons. He published his theory in 1920, with little proof. It was rejected and even ridiculed. At the time, chemists believed that polymers were small molecules held together by other forces.

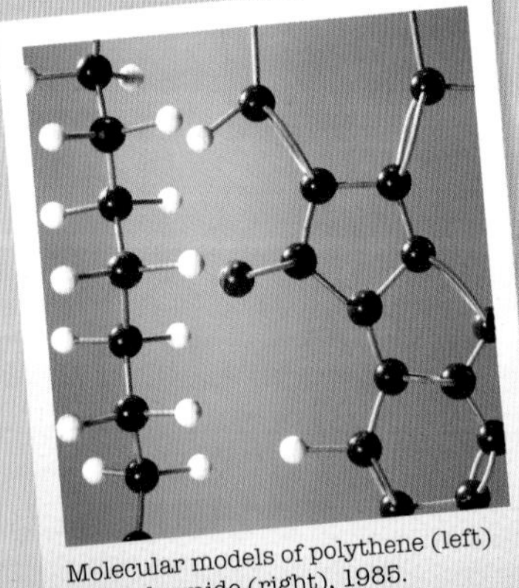

Molecular models of polythene (left) and polyamide (right), 1985.

1941
THE HAPPY ACCIDENT OF NON-STICK

Random chance has triggered many scientific discoveries, among them the development of the non-stick coating, Teflon. Roy Plunkett was working on synthesizing refrigerants at DuPont. One of his prepared cylinders of gas did not release gas but was too heavy to be empty. Cutting it open, he found the inside coated with a waxy, white powder. It had a very high melting point, was completely unreactive and was more slippery than any previously known substance. Plunkett realized that the high pressure in the cylinder had forced the gas molecules to join together, forming a new polymer. He abandoned his other work and focused on Teflon, which DuPont marketed in 1941. Its uses now extend from non-stick cookware to coating heat shields on spacecraft.

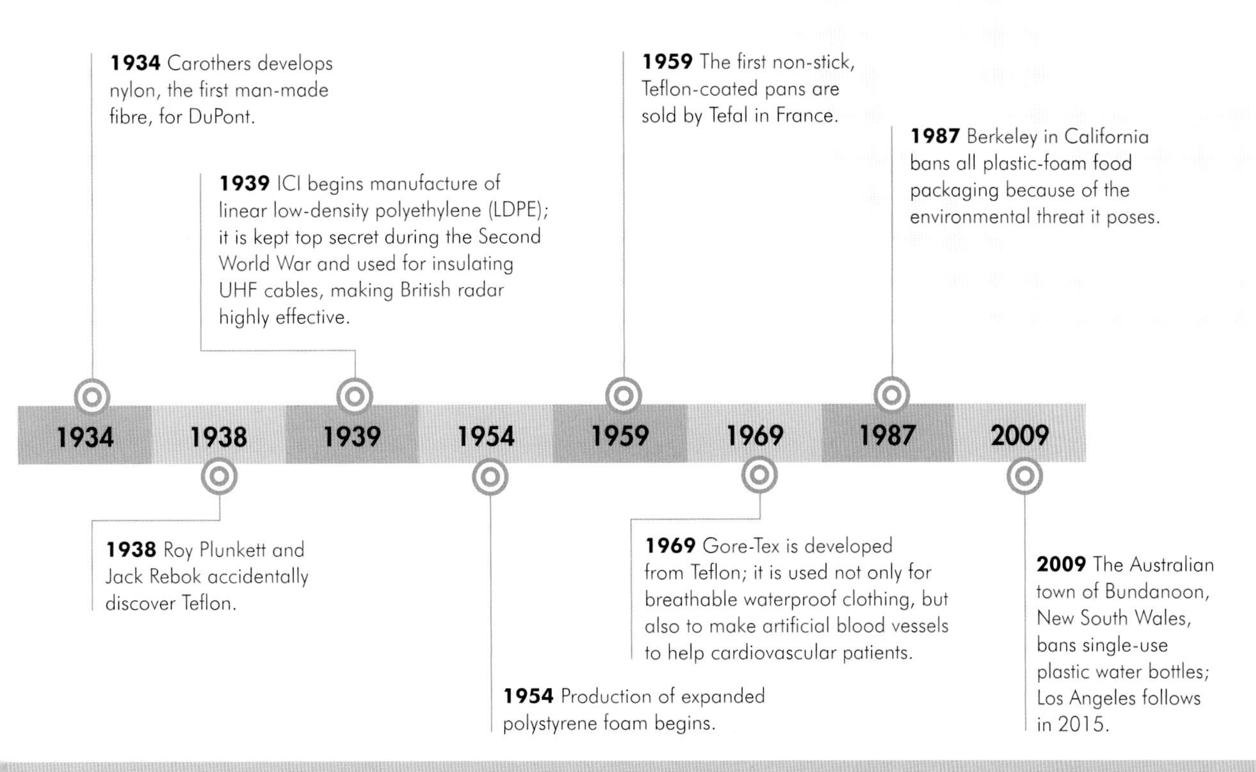

1930s–1950s
SYNTHETICS EVERYWHERE

From the 1930s to the 1950s, more synthetlcs were discovered and more variants of them developed. The most important were the plastics based on polythene. Plastic items were initially considered exotic, but very soon became ubiquitous. Their use extended from durable items intended for a long life to disposable items and packaging. By the end of the century most foods and household items came packed in plastic of one kind or another, often in the form of thin plastic film. Shoppers picked up plastic carrier bags with every purchase, and most fabrics included at least some portion of artificial polymer fibres. Polymers have made consumer goods affordable for millions, including those in developing economies.

Post 1931
THE BACKLASH

There is a massive disadvantage to the unreactive nature of artificial polymers: they are hard to destroy. While natural organic materials such as wood and cotton are quite easily broken down, plastics, nylon, acrylics and other oil-based polymers are non-biodegradable. Plastics began to swamp landfill sites and fill the seas. The plastic carrier bag used for an hour or two will take thousands of years to break down. The garbage patches of the Pacific Ocean contain around 100 million tonnes of floating plastic debris. Attempts to tackle the problem include recycling, making biodegradable plastics from plant material (cellulose) and banning or limiting plastic products.

Tupperware parties became a successful way of selling plastic food-storage containers.

The proliferation of plastic goods in landfill is a serious environmental problem.

1931
THE BIG BANG

'As far as I can see, such a theory remains entirely outside any metaphysical or religious question. It leaves the materialist free to deny any transcendental Being...For the believer, it removes any attempt at familiarity with God... It is consonant with Isaiah speaking of the hidden God, hidden even in the beginning of the universe.'

– Georges Lemaître

Diagram explaining the Big Bang theory, 1933.

Most people now believe that the universe started with a bang – a big one. Yet the idea was scorned when first suggested, even by Einstein whose theory of relativity supported it.

Alexander Friedmann in Russia and Georges Lemaître in Belgium arrived independently at the idea of an expanding universe, both working from Einstein's relativity equations. Friedmann modelled an expanding universe only as a mathematical exercise; he also modelled a steady-state universe. Lemaître went further, marshalling cosmological observations and proposing the expanding universe as a theory in 1927. Examining Hubble's work on other galaxies in 1924, Lemaître concluded that if galaxies are receding in all directions, the universe must indeed be expanding. His work went largely unnoticed until 1930, when Arthur Eddington drew the attention of other cosmologists to it and arranged its translation into English. Einstein later called his scepticism about Lemaître's theory 'the biggest blunder of [his] life'.

In 1930, Lemaître extended his theory to its logical conclusion. If the universe is expanding, it must have been much smaller previously – it follows that it was once a tiny point that burst into our current universe at some very distant point in time. Lemaître referred to this original point as a 'primaeval atom' or 'cosmic egg', though it is now known as a 'singularity'.

FLASHPOINTS FACT

The name 'Big Bang' was coined in 1950 by astronomer Fred Hoyle, one of the theory's detractors, as a term of ridicule.

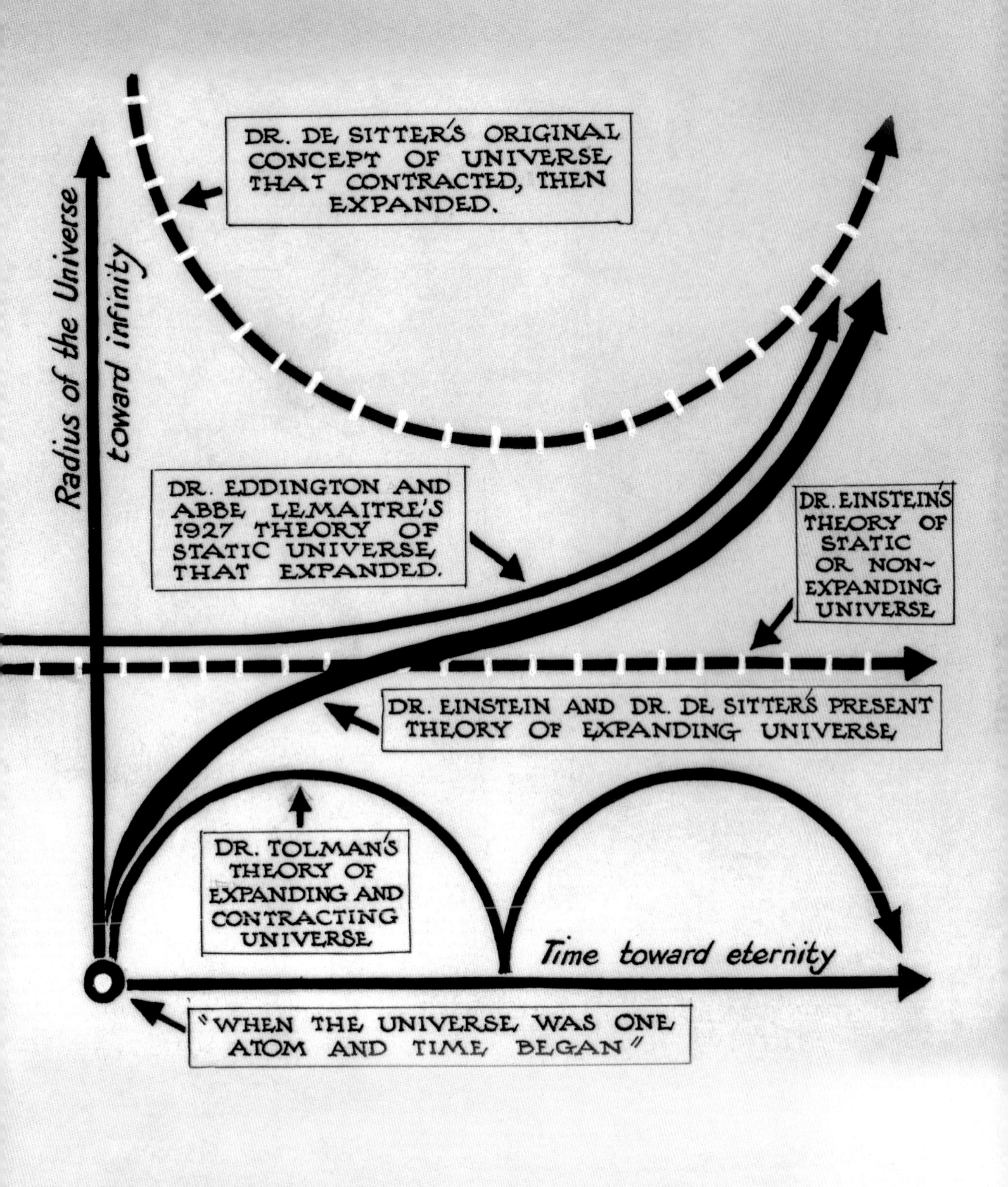
DR. DE SITTER'S ORIGINAL CONCEPT OF UNIVERSE THAT CONTRACTED, THEN EXPANDED.
Radius of the Univers
toward infinity
DR. EDDINGTON AND ABBE LEMAITRE'S 1927 THEORY OF STATIC UNIVERSE THAT EXPANDED.
DR. EINSTEIN'S THEORY OF STATIC OR NON-EXPANDING UNIVERSE
DR. EINSTEIN AND DR. DE SITTER'S PRESENT THEORY OF EXPANDING UNIVERSE
DR. TOLMAN'S THEORY OF EXPANDING AND CONTRACTING UNIVERSE
Time toward eternity
"WHEN THE UNIVERSE WAS ONE ATOM AND TIME BEGAN"

TIMELINE

1912 Vesto Slipher finds spiral nebulae are all moving away from Earth, but does not work out the cosmological implications.

1922 Alexander Friedmann produces a mathematical model of an expanding universe.

1924 Edwin Hubble establishes the existence of other galaxies outside the Milky Way.

1927 Lemaître publishes his work on the expanding-universe theory; it goes largely unnoticed.

1930 Lemaître's work is translated into English and attracts a great deal more attention and support.

1931 Lemaître publishes his theory that the universe expanded from a single point.

FLASHPOINTS

1225
INFINITY AND BEYOND

Nearly 2,500 years ago, Aristotle proposed that the universe had an infinite past. The idea is not just difficult to grasp, but impossible to reconcile with most religious models. In 1225, Robert Grosseteste suggested that the universe began with an explosion and the subsequent crystallization of matter produced the stars and planets. For the early part of the modern era, though, there was general agreement that the universe was unchanging – a solid-state model that many cosmologists clung to until the 1960s. The groundwork for an expanding-universe theory was laid in the second decade of the 20th century.

Lemaître's Big Bang theory has the universe expanding from a single 'atom', now called a singularity.

1912
GALAXIES HERE, THERE AND EVERYWHERE

In 1912, the astronomer Vesto Slipher measured the Doppler shift (red shift) of spiral nebulae (galaxies) and found that they are all moving away from Earth. His discovery had little impact and no one really knew what the nebulae were anyway; Slipher did not work out the cosmological implications of his discovery. Twelve years later, Edwin Hubble, using the most powerful telescope of the time, discovered that the nebulae are in fact other galaxies. Lemaître put together these two pieces of information, along with Einstein's equations, and concluded that the galaxies are being pushed away from Earth because the space between them and us is expanding.

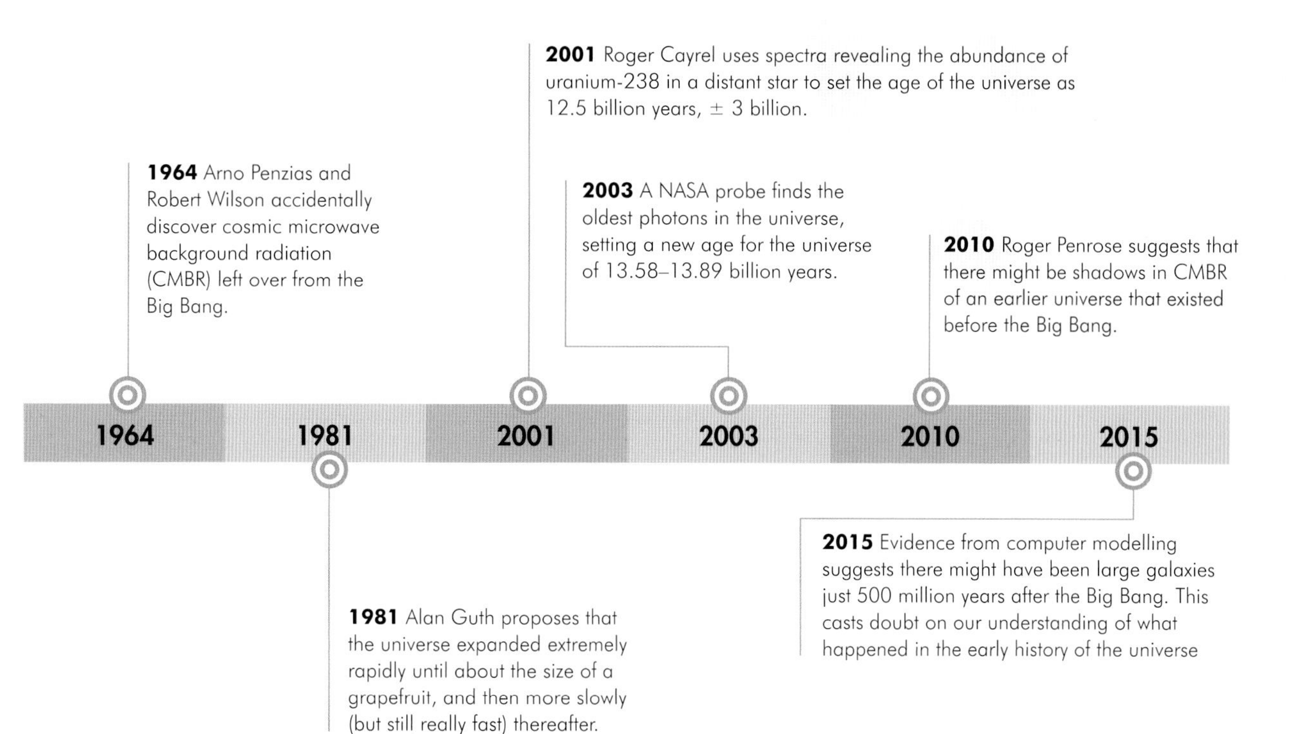

Post 1931
BIG BANG MAKES A BIG SPLASH

Although some cosmologists accepted Lemaître's model, many did not, and some suggested conflicting interpretations of Hubble's data. Some favoured an oscillating universe that repeatedly expands and then contracts and then expands again in a cycle of bangs and crunches. Fritz Zwicky even proposed that red shift was produced by photons getting tired and losing energy with age. In the model of an expanding universe, objects stay the same size, but the space between them increases – rather as points on the surface of a balloon move further apart as the balloon is inflated.

1964
BACK TO THE START OF SPACE-TIME

The dispute between those who favoured a solid-state model and those who favoured expansion or other models rumbled on until 1964. Then Arno Penzias and Robert Wilson found the 'echo' of the Big Bang. They were trying to eliminate radio noise interfering with their radio telescope observations. After several attempts, including shooting pigeons they suspected of disturbing their apparatus, they discovered that the source was CMBR – leftover thermal radiation from the Big Bang. Now only the most stubbornly intransigent cosmologist could favour a different theory.

Physicists have since modelled the early universe, going back as far as the first 0.0000000000 0000000000000000000000000000001 second (the end of the Planck era, when space-time formed).

The ECHO horn radio antenna with which Wilson and Penzias discovered cosmic microwave background radiation left over from the Big Bang.

1931

THE ELECTRON MICROSCOPE AND THE DAWN OF MICROBIOLOGY

'The Microbe is so very small
You cannot make him out at all,
But many sanguine people hope
To see him down a microscope.'

– Hilaire Belloc, 'The Microbe' in *More Beasts for Worse Children*

Electron microscopes enabled scientists to examine smaller items than had ever been seen before.

When optical microscopes first appeared in the early 1600s, they opened up a world of unimagined detail, including the fine structures of our own bodies and micro-organisms that swarmed in the air and water. By the end of the 19th century, though, optical microscopes (those that work with light) had reached the limits of their capabilities.

The first step towards electron microscopy came in 1924, when Louis de Broglie showed that electrons could behave like waves. This meant they could be used in the same way as light, but with a much higher resolution as their wavelength is much smaller than that of visible light. In 1931 Ernst Ruska and Max Knoll, working on oscilloscopes, discovered that a beam of electrons could be focused using an electromagnet. With two beams, they could make a simple electron microscope. This early attempt achieved only a low resolution and had a tendency to set fire to the samples, but was an important proof of concept. The first commercial electron microscope, in 1939, had a resolution of ten nanometres.

The electron microscope has made it possible to see the structure of viruses, and the molecular structure of proteins and other large molecules. There are limitations, though, as the technique involves the destruction of, or damage to, some types of samples, including biological samples and the components of semiconductors.

The environmental scanning electron microscope (ESEM), developed in the 1980s, addressed some of the problems of sample destruction and the limitations on use. It has made electron microscopy more versatile. But the very latest electron microscopes are even more stunning. They can work with objects a million times smaller than the width of a human hair, enabling scientists to witness individual atoms for the first time.

FLASHPOINT FACT

In 1990, scientists at IBM used an electron tunnelling microscope to spell out the letter 'IBM', using 35 xenon atoms on chilled nickel crystal. It was the first time individual atoms had been manipulated.

TIMELINE

1590s The optical microscope is invented, probably in the Netherlands.

1676 Antonie van Leeuwenhoek reports the discovery of micro-organisms with his microscope.

1897 J.J. Thomson discovers the electron, the first subatomic particle to be found.

1903 Richard Zsigmondy develops the ultramicroscope, able to show objects smaller than the wavelength of light. It uses light scattering, rather than reflection, and can only show a fuzzy image.

1924 Louis de Broglie shows that electrons can behave like waves as well as particles.

1931 Ernst Ruska and Max Knoll begin development of the first electron microscope.

FLASHPOINTS

1676
THE WORLD IN MICRO

At some time in the 1590s, a scientist in the Netherlands first saw some of the microstructure of the natural world. The invention of the first microscope is not recorded, but it was soon copied. In 1676, Antonie van Leeuwenhoek reported seeing micro-organisms – tiny living things – that swarmed in pond water and even the human mouth. Van Leeuwenhoek made many microscopes and explored the structure of natural objects he found around him. He was the first to see sperm, to detail the structure of muscle fibres and to see bacteria.

Anton van Leeuwenhoek.

1924
LIGHT IS NOT GOOD ENOUGH

Optical microscopes (those that work with light) produce magnified images using a series of mirrors and/or lenses. For many years, the limit on their power was the quality of lenses that could be ground, but eventually they reached the limit of the wavelength of visible light, around 250 nanometres, at around magnification 1,000x. This was a limit imposed by physics itself – the nature of light and its ability to resolve smaller detail.

The only way to progress beyond this barrier was to work with something that had a smaller wavelength than visible light. When Louis de Broglie showed that electrons could behave as a wave, they offered a way forward.

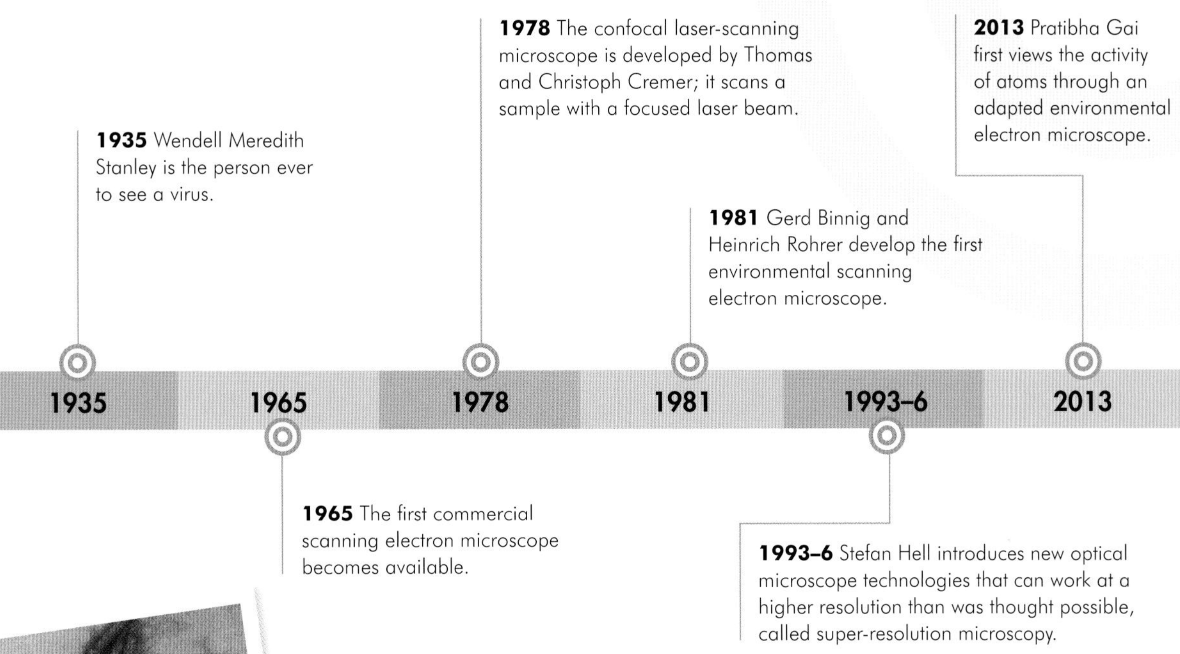

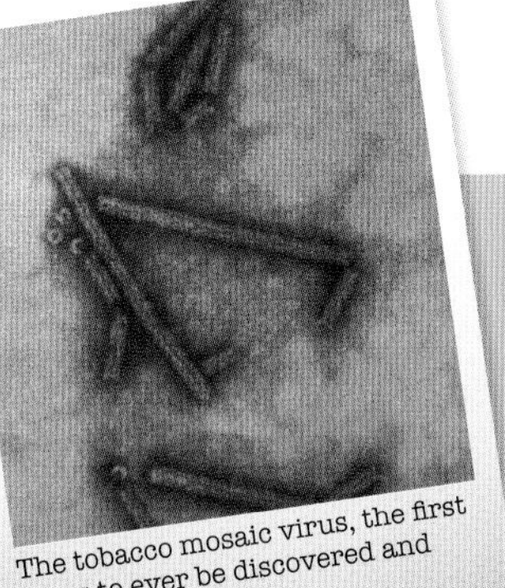
The tobacco mosaic virus, the first virus to ever be discovered and key to gene therapy.

1935
WHAT CAN YOU SEE?

The electron microscope revealed viruses for the first time. Much smaller than bacteria, their existence had been demonstrated before they were actually visible. A virus is little more than a string of DNA or RNA with a protein coat – many can fit inside a single human or animal cell. They range from around 20 to 400 nanometres in diameter. The first virus ever seen was the tobacco mosaic virus, isolated and viewed by Wendell Meredith Stanley in 1935. Modern electron microscopes offer magnification up to 2,000,000x and can resolve large molecules.

1965
STILL GOING

The original electron microscope was a transmission electron microscope (TEM), which creates an image from a beam of electrons that is in part scattered by the sample and in part passes through it. The scanning electron microscope (SEM) scans the sample with a beam of electrons and measures the absorption of their energy by different methods. Work on an SEM was begun by Manfred von Ardenne, but his prototype was destroyed in a bombing raid during the Second World War. Charles Oatley continued the work in Britain after the war and the first working SEM was built in 1965. This used a different technique, scanning the sample and measuring the ways energy is absorbed.

1933
RADIO FROM THE STARS

'Radio astronomy has, in the last decade, opened a new window on the physical universe. It may also, if we are wise enough to make the effort, cast a brilliant light on the biological universe.'

– Carl Sagan, 1978

Grote Reber working with his radio telescopy equipment, 1948.

Visible light forms only a tiny portion of the electromagnetic spectrum. As we see light beamed through space every day, it is hardly surprising that other types of electromagnetic radiation also reach us, though it was not thought possible at first.

In 1931, Karl Jansky was investigating interference on long-distance voice transmissions for Bell Telephone Laboratories when he found a persistent hiss that was most pronounced at about the same time each day. He ruled out thunderstorms and even the sun, as he found that the repeat cycle was 23 hours and 56 minutes – the same as Earth's rotation relative to the stars. Recognizing that the signal came from space, he consulted star maps and identified the source as somewhere near the centre of the galaxy. He announced his astonishing discovery in 1933, but Bell refused his request for funding to build a radio telescope.

Four years later in September 1973, Illinois-born Grote Reber built the first radio telescope in his back garden. It was made up of a parabolic sheet-metal mirror 9m (30ft) in diameter, focusing to a radio receiver 8m (26ft) above the mirror. The telescope was mounted on a tilting stand which allowed it to be pointed in different directions, although not actually turned. It took him three attempts, but eventually Reber managed to construct a telescope that could detect signals from outer space.

Ever since then, radio telescopes have profoundly changed what we know about the universe. They focus incoming electromagnetic radiation on an antenna, then convert the signal to a tiny electric current, which is recorded. The strength and frequency of the signal are measured. If or when we discover life elsewhere in the universe, it will be a radio telescope that finds the signal.

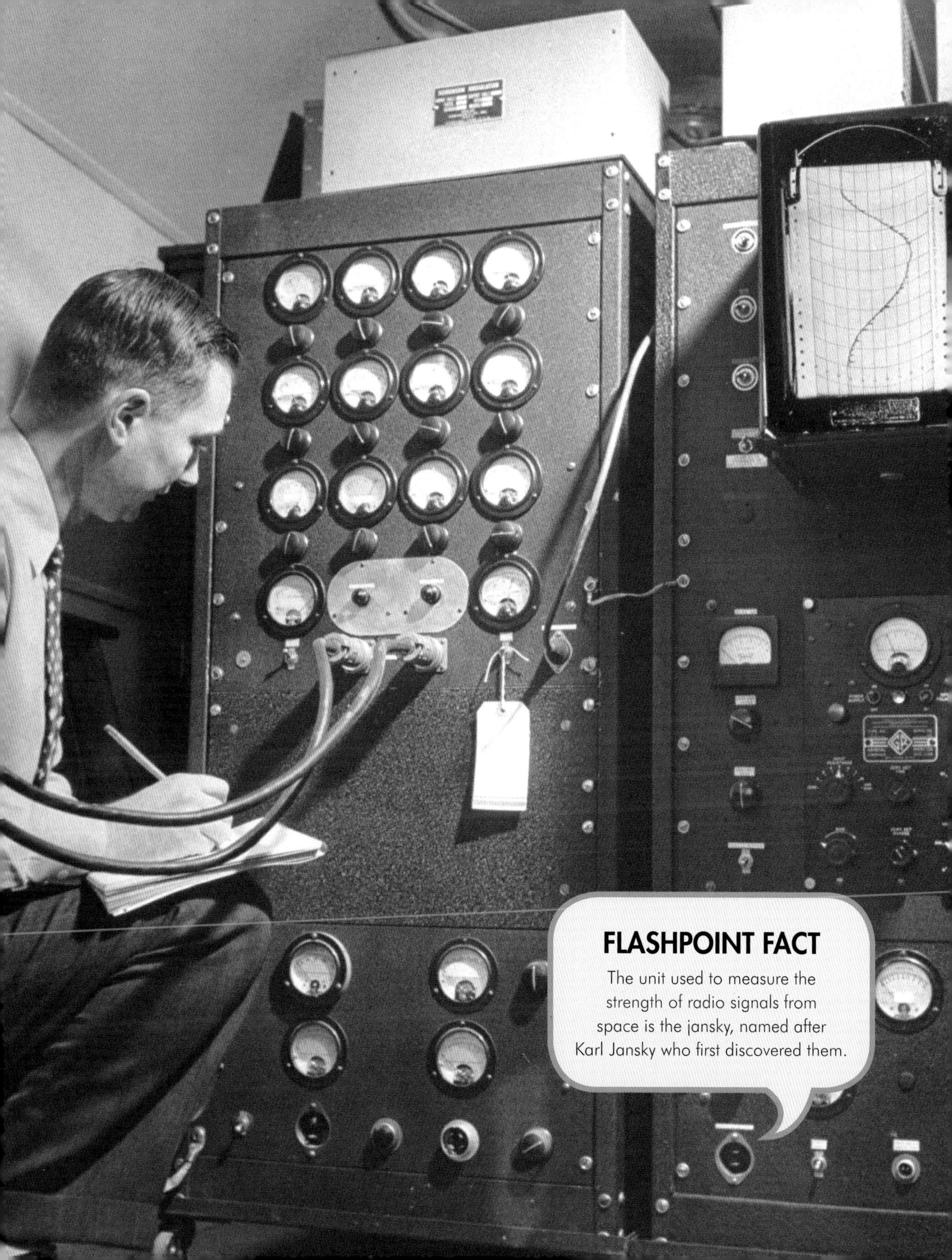

FLASHPOINT FACT

The unit used to measure the strength of radio signals from space is the jansky, named after Karl Jansky who first discovered them.

TIMELINE

1608 | 1609 | 1896 | 1902 | 1932 | 1937 | 1957

1608 The first optical telescopes (those that work with light) are made in the Netherlands.

1609 Galileo uses a telescope to see details of the moon and planets for the first time.

1896 Johannes Wilsing and Julius Scheiner attempt to pick up radio emissions from the sun, but fail.

1902 Physicists conclude that radio waves from space would bounce back off the Earth's atmosphere and be undetectable.

1932 Karl Jansky realizes that radio interference on voice signals is coming from distant space.

1937 Grote Reber builds a 9-m (30-ft) parabolic dish radio telescope in his back garden.

1957 The massive radio telescope at Jodrell Bank, UK, is built to investigate cosmic rays. It can be tilted and directed towards any part of the sky. At 80m (262ft) across, it is still the third-largest steerable telescope in the world.

FLASHPOINTS

1609
SECRETS OF THE PLANETS REVEALED

The first time the dots of light in the night sky resolved into anything else was in 1609, when Galileo built the first telescope capable of revealing the detail of the planets. Galileo had heard of the invention of the telescope in Flanders in 1608 and rushed to make his own, better, model. Instead of 3x magnification, Galileo's telescope had 20x magnification. With it, he could see the surface of the moon and – more importantly – the rings of Saturn and the moons of Jupiter, and that the planets were physical bodies, perhaps like the Earth. It was the first step towards exploring space.

Galileo demonstrating his telescope in Venice.

1963
BIGGER AND BETTER

Large radio telescopes can be single dishes or many dishes and antennae linked together (array telescopes). The Arecibo radio telescope in Puerto Rico has a dish 305m (1,000ft) across. Since its opening in 1963, it has been instrumental in discovering the first evidence of neutron stars, binary pulsars and exoplanets (planets outside our solar system). It is also the base of SETI, the Search for Extra-Terrestrial Intelligence, hunting for evidence of alien intelligence. Array telescopes are used for inferometry, which adds together signals picked up by many antennae, sometimes spread over a large area. They are best for resolving fine detail from bright objects. The Karl G. Jansky Very Large Array in New Mexico has 27 telescopes over a span of 80km (49 miles).

Karl Guthe Jansky with his directional radio aerial system.

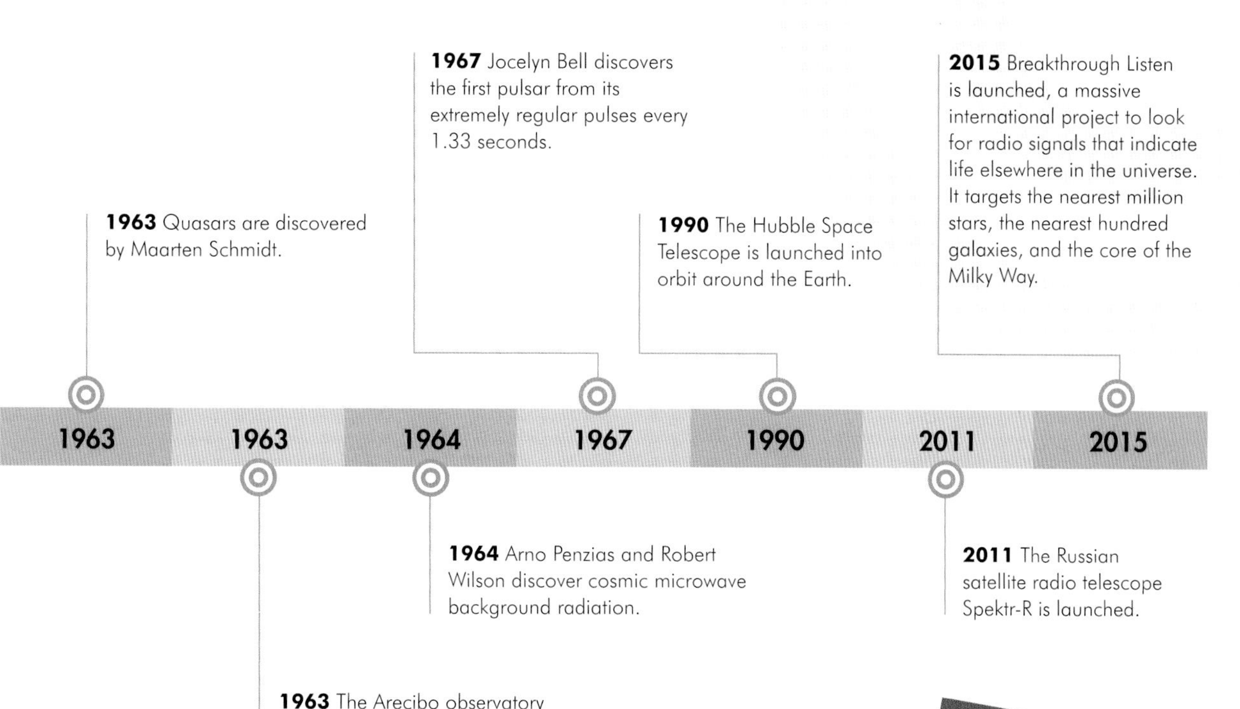

1967
RADIO FROM SPACE

Radio signals from space have numerous sources, including supernovae, pulsars, quasars and galaxy remnants. Monitoring emissions from sources within the solar system enables astonomers to calculate, for example, the temperatures of other planets. Monitoring radio from deep space has led to many discoveries. When Jocelyn Bell discovered a highly regular pulse of radio signals in 1967, she at first suspected that it was a deliberate artificial signal that could be a sign of intelligent life. In fact, it turned out to be the first pulsar – a highly magnetized rotating neutron star.

1990
MOVING INTO SPACE

Both optical and radio telescopes suffer from waves being bent or reflected by the Earth's atmosphere, but telescopes sited on satellites above the atmosphere don't suffer from this problem. The Hubble Space Telescope, a high-resolution optical telescope, was launched in 1990 and has provided thousands of stunning photographs of distant phenomena as well as objects within the solar system. The Russian space-based radio telescope Spektr-R (launched in 2011) is used in conjunction with radio telescopes on Earth using inferometry techniques. Other space-based telescopes operate at the other end of the spectrum, detecting gamma rays or X-rays.

Engineer Grote Reber installing a radiometer dish, which is used to study the effects of solar waves on radio transmissions, 1948.

1936

TURING: TOWARDS THINKING MACHINES

'Turing's paper…contains, in essence, the invention of the modern computer and some of the programming techniques that accompanied it.'

– Marvin Minsky, 1967

The Bombe, rebuilt at Bletchley Park, was used to crack German codes during the Second World War.

Alan Turing (1912–54) is in many senses the father of modern computing. He proposed a theoretical model for a programmable computer in 1936, which has since come to be seen as the foundation of subsequent actual computers. Now called the 'Turing machine', his idea described a machine capable of performing any calculations by following a series of logical instructions.

During the Second World War, Turing worked at Bletchley Park, the codebreaking centre in England. He was instrumental in developing technology to crack the coding of a German encryption machine known as Enigma. The de-encryption machine, the 'bombe', is credited with shortening the war by up to four years. After the war, Turing worked at the National Physical Laboratory, where he designed ACE, one of the first stored-program digital computers. In 1948 he moved to Manchester University, where pioneering work in computing was being carried out.

There, in 1950, he published a groundbreaking paper, 'Computing Machinery and Intelligence', in which he suggested the highly provocative idea that a computer could be designed that could learn and then think independently. He formulated a test for such a computer, which a computer could pass if it were able to fool a human interrogator into believing it was human. This test, known as the 'Turing Test', has been explored extensively since 1950 and is still a subject of popular debate today.

FLASHPOINT FACT

Alan Turing killed himself by eating an apple he had injected with cyanide. He had a great liking for the Disney movie *Snow White*, and it has been suggested that he took the idea of the poisoned apple from the film.

TIMELINE

1770 Wolfgang von Kempelen introduces his chess-playing automaton, the Mechanical Turk – but it is a hoax, with a human chess player hidden inside.

1912 Alan Turing is born in London.

1936 Turing suggests his theoretical model for an a-machine (automatic machine) and for a universal Turing machine.

1943 Colossus, the first electric, digital, programmable computer, is demonstrated at Bletchley Park; it is used to crack the German Lorenz cipher and is kept secret.

1950 Alan Turing publishes 'Computing Machinery and Intelligence'.

1951 Dietrich Prinz develops the first computer chess program, though it can only solve short chess problems rather than play a whole game.

1954 Turing takes a lethal dose of cyanide after prosecution and punishment for homosexual acts.

FLASHPOINTS

1936
THE UNIVERSAL TURING MACHINE

In 1936, Turing described an 'a-machine' (or 'automatic machine'), a hypothetical machine that acts as a blueprint for any computer. It would read data from a paper tape and manipulate it following a set of instructions (a program). Turing proposed that given enough memory and time, any computational task could be carried out – it just had to be expressed as rules (algorithms). The combination of memory, program, processing, input and output has come to define all subsequent computers.

He also proposed that a powerful enough a-machine (now called a 'universal Turing machine') would be able to emulate any other computer as it just needs to follow instructions to behave in the same way. This was revolutionary at a time when machines were designed to tackle a single task.

The Pilot Ace computer, designed by Alan Turing.

1950 and 1966
THE TURING TEST AND ELIZA

Turing's challenge for computer intelligence has become known as the 'Turing Test'. It requires a human interrogator to question, by text, two other 'subjects', one a human and the other a computer. The computer passes the test if the interrogator cannot identify the computer 70 per cent of the time after a five-minute 'conversation'.

An early natural-language program, ELIZA, written by Joseph Weizenbaum in 1966, examined patterns in speech in order to respond to human remarks with apparently pertinent questions. One version quite successfully imitated a therapist, responding to such remarks as 'My mother hates me' with, 'Who else in your family hates you?' A number of people were fooled into believing ELIZA showed genuine intelligence.

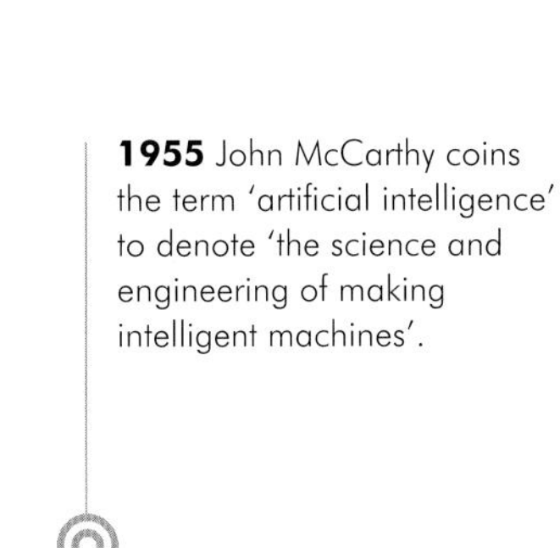

1955 1958 1966 1974 1997 2011

1955 John McCarthy coins the term 'artificial intelligence' to denote 'the science and engineering of making intelligent machines'.

1958 The chess program NSS defeats a novice human chess player, the first time a computer outwits a human being.

1966 Joseph Weizenbaum creates ELIZA, the first natural-language processing program that can hold a type of conversation with humans.

1974 The existence of Colossus is finally revealed, 30 years later.

1997 Chess grand master Garry Kasparov is beaten by an IBM computer playing chess.

2011 Apple introduces Siri, the 'intelligent personal assistant', as part of iOS5 on iPhones.

1955
DEFINING INTELLIGENCE

The field of artificial intelligence (AI) was first defined by John McCarthy in 1955. One problem that besets research and development in AI is that there is no agreed definition of intelligence of any type. Theories of human and animal intelligence prioritize different aspects. Which definition of intelligence is to be emulated? Further, there is no single approach to developing machine intelligence, with some researchers wishing to emulate human intelligence and working from neurology and psychology, while others look for completely novel ways of constructing intelligence. As they point out, we don't model planes on the way birds fly, so why should we try to copy the working of the human brain to produce intelligence?

1997
GETTING SMARTER

Games were one early approach to intelligence, and even Turing had played around writing a computer program to play chess (though only on paper). Now, the most sophisticated chess programs can beat any human – that barrier was broken in 1997, when IBM's Deep Blue defeated the reigning world chess-champion, Gary Kasparov. But Deep Blue is more an 'expert system' than genuinely intelligent. Expert systems search and collate a massive dataset, try out many solutions very rapidly, and some 'learn' from past mistakes, but they can't go beyond their initial programming and knowledge base.

World champion Garry Kasparov plays chess against the IBM chess program, Deep Blue, in 1997.

1937

THE KREBS CYCLE

'Both the body and its parts are in a continuous state of dissolution and nourishment, so they are inevitably undergoing permanent change.'

– Ibn al-Nafis, *The Treatise of Kamil on the Prophet's Biography*, 1260

A close-up of mitochondria, which generate the chemical energy that cells need to do their various jobs.

One of the most crucial discoveries about how the human body works has to be the mechanism for producing energy from food. This, at the cellular level, is what drives life. The pieces of the puzzle were put together in 1937 by Hans Krebs. The second metabolic cycle to be discovered by Krebs, the citric acid cycle – or Krebs cycle – outlines the stages of the process by which cells break down glucose to provide energy, known as cellular respiration.

There are many stages to the Krebs cycle, which starts when a glucose molecule $C_6H_{12}O_6$ is broken down into two pyruvate molecules, $C_3H_4O_3$. The pyruvate molecules are broken down over two circuits of the Krebs cycle, generating adenosine triphosphate (ATP), which stores chemical energy to be used in other cell processes, and CO_2 (amongst other products). The cycle takes place in the mitochondria inside cells.

All aerobic (oxygen-requiring) animals use the Krebs cycle as their means of breaking down glucose from food (or from their own production during photosynthesis, in the case of plants). It is thought to be one of the earliest metabolic pathways to have evolved, and is crucial to life on Earth. It was one of three pathways discovered by Krebs.

Many more metabolic pathways have since been discovered. Together they explain the entire working of living organisms in terms of the biochemistry of cells. They involve enzymes and chemicals taken in through feeding, respiration and photosynthesis and cover metabolic processes from digestion to growth and reproduction.

DID YOU KNOW?

When Krebs submitted his short paper on the citric acid cycle to the prestigious journal *Nature*, it was rejected due to lack of space. In 1953, he won a Nobel Prize for the work described in that paper.

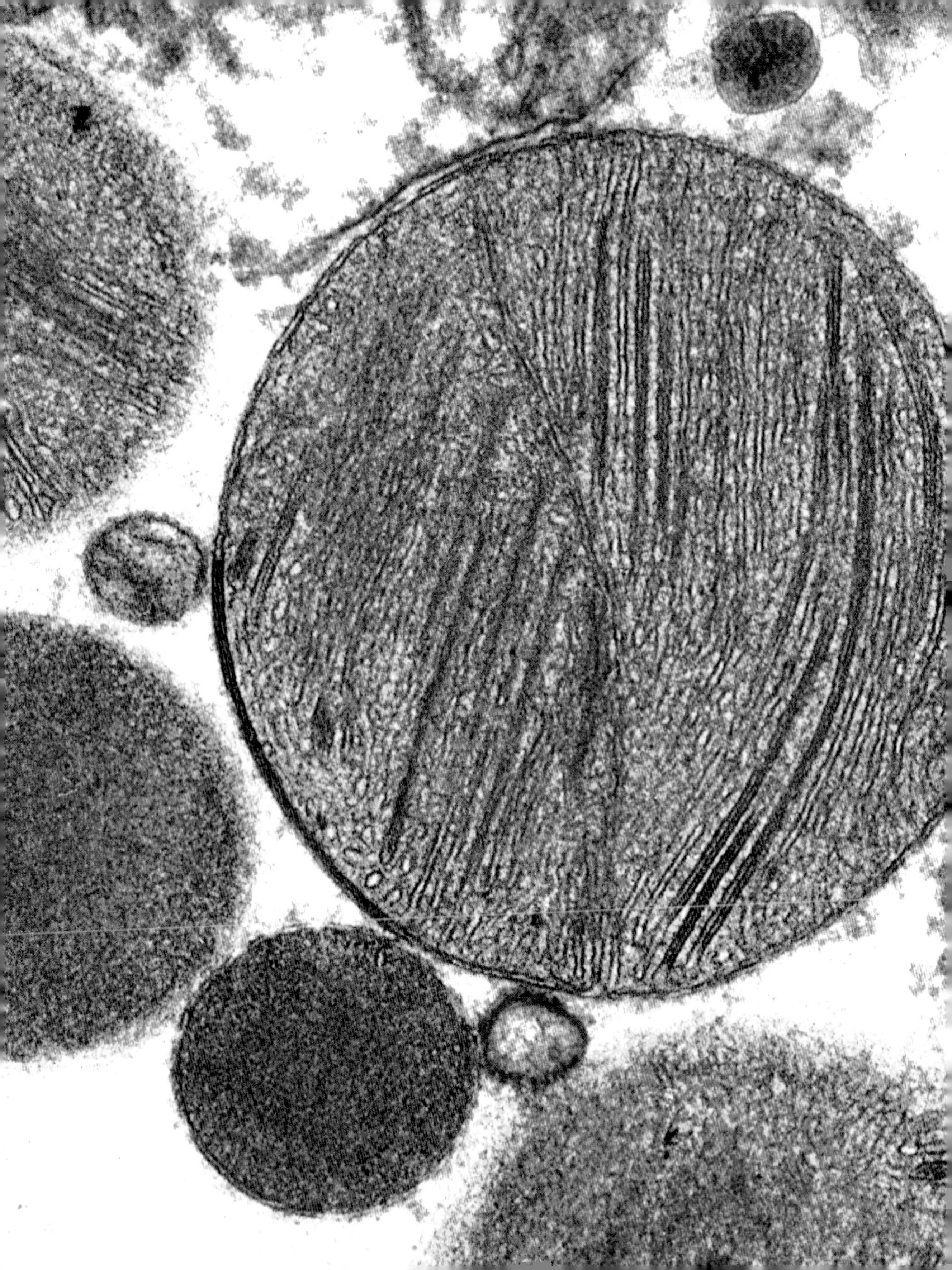

TIMELINE

1784 Antoine Lavoisier and Pierre-Simon Laplace demonstrate that we use oxygen and produce carbon dioxide and water during respiration.

1805 John Dalton publishes his theory of the atomic makeup of all matter.

1816 François Magendie reports his findings that dogs cannot take nitrogen from the air and must take it from food.

1828 Friedrich Wöhler discovers that chemical synthesis and reactions in cells in the body are no different from the same phenomena anywhere else.

1842 Justus von Liebig declares that the energy needed by muscles comes entirely from breaking down protein, with urea as a byproduct.

FLASHPOINTS

1784
HOW BODIES WORK

The groundwork for recognizing and exploring metabolic pathways was laid over centuries of research into physiology, nutrition and chemistry. In 1784, French chemists Antoine Lavoisier and Pierre-Simon Laplace studied respiration and discovered that the body takes in oxygen and produces carbon dioxide. In 1816, François Magendie reported his findings from experiments with nutrition in dogs. He had fed dogs a single type of food – first sugar, later olive oil, butter, gum, bread and other substances. In all cases the dogs sickened and died, showing – he said – that they can't take nitrogen from the atmosphere but must take it from food.

Molecular model of citric acid, made of hydrogen (white), carbon (grey) and oxygen (red).

1805
WE ARE CHEMICAL

In 1805, John Dalton published his theory that all matter is made of atoms and all chemical reactions involve the rearrangement of atoms in different compounds. This fundamental discovery could explain how we could make our bodies and derive energy from food. All that remained was to work out the nitty-gritty of how it happened. It turned out to be far more complicated than anyone could have begun to imagine. The first person to begin investigating how organisms process chemicals was Louis Pasteur, who concluded in 1857 that fermentation does not happen when yeast cells die, but that alcohol is a product of their normal metabolism while living.

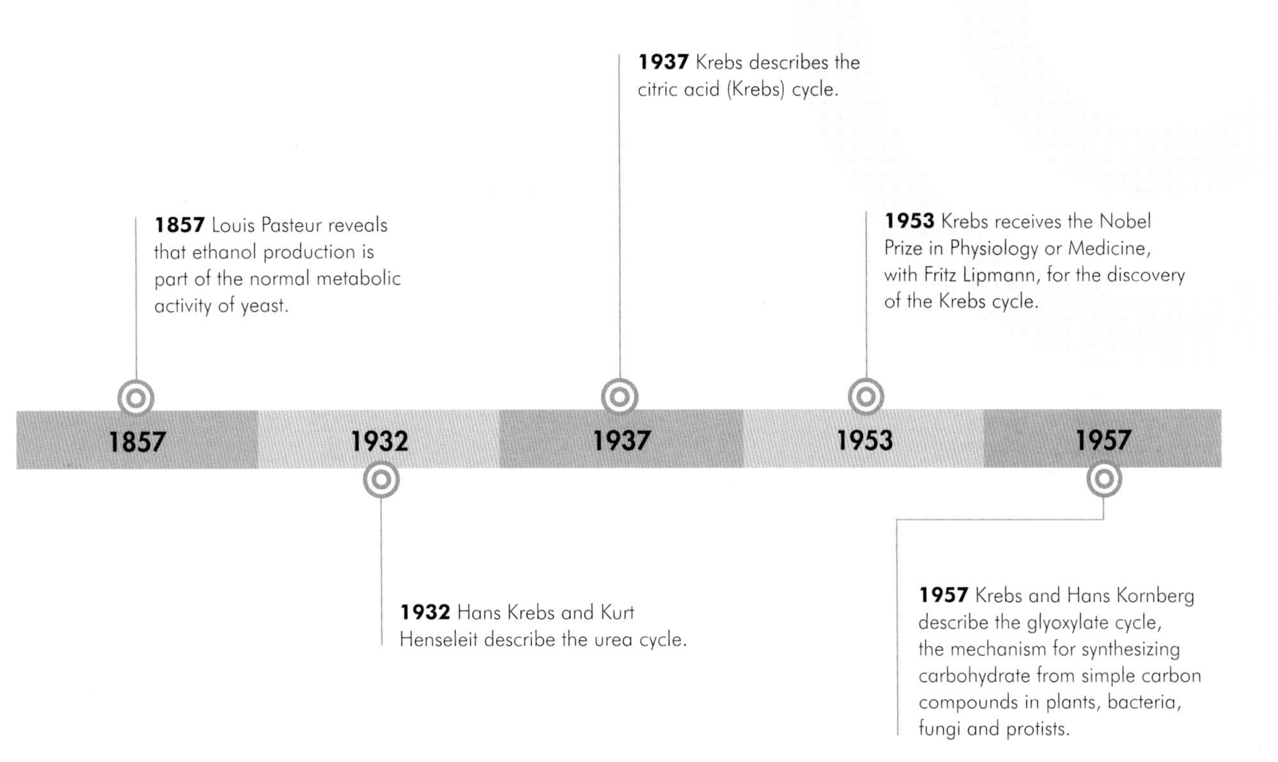

1921
CYCLING ON

There are many metabolic pathways and cycles that define the ways our bodies use chemicals in the series of reactions that keep life going and bodies growing. The recognition that all the processes of life are governed by biochemical patterns and responses has led to the identification of a large number of medical conditions that arise when biochemistry goes wrong. Liver failure leads to disruption of the urea cycle and can be diagnosed because there is excess ammonia in the blood. Diabetes, as discovered in 1921, can be successfully treated with insulin, because insulin triggers cells to convert sugar into storable sources of energy.

1932
KREBS AND THE UREA CYCLE

The first of the human metabolic cycles to be discovered was the urea cycle, described by Krebs working with Kurt Henseleit in 1932. Ammonia is produced by the breakdown of protein in the gut; it is highly toxic. The body converts it to urea, or uric acid, to get rid of it in the urine. Most of the work is done in the liver in mammals, including humans, with some taking place in the kidneys. Failure of the urea cycle causes serious illness and death, as a build-up of ammonia can lead to liver failure. The urea cycle consists of five reactions that take place in turn, beginning with ammonium and carbonate ions and producing as final end products urea and an amino acid, ornithine.

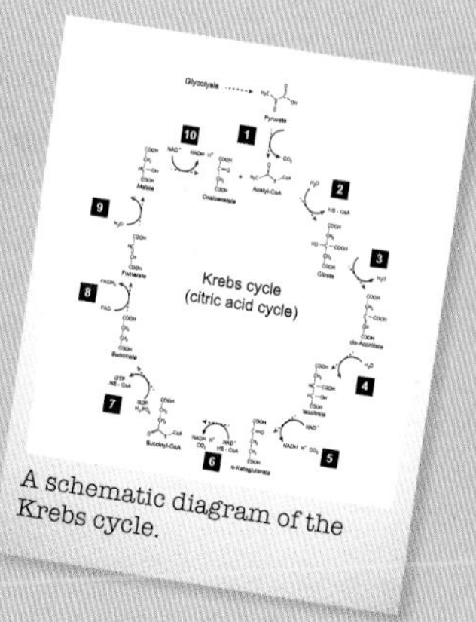

A schematic diagram of the Krebs cycle.

1940–1959

The 1940s saw another world war, another disruption to scientific progress in many areas – yet it was a spur to certain aspects of physics. Rocketry, which was just beginning before the war, was developed rapidly by German scientists to carry bombs around Europe. Nuclear fission became the focus of attention in the USA as the Allies rushed to develop an atomic bomb ahead of the Germans. After the war, these developments were put to more peaceful uses – the beginnings of space travel and the start of nuclear power. Meanwhile, many of the physicists who had become disenchanted with the subject after seeing the destruction wrought by the atom bomb turned to biology and biochemistry. The 1950s saw advances in medicine and genetics, with the first successful transplant operations, the development of effective vaccines against polio and other diseases and the commercial development of antibiotics. It was a decade dedicated to saving lives, after the carnage of the Second World War.

Not everything was peaceful, however. The increasing hostility between the USA and the USSR crystallized into the Cold War, which dominated western politics until the 1990s. It, too, was a spur to scientific advances, though. The space race was a direct consequence of the competition between the two superpowers – and the 1950s ended with the USSR well ahead.

1942
THE NUCLEAR CHAIN REACTION

'It does not seem possible, at least in the near future, to find a way to release these dreadful amounts of energy – which is all to the good, because the first effect of an explosion of such a dreadful amount of energy would be to smash into smithereens the physicist who had the misfortune to find a way to do it.'

– Enrico Fermi, 1923

The atomic bomb explodes in Hiroshima, viewed from Kure, Japan.

It's not always possible to tell whether a development or discovery will be put to good or bad uses, but the realization that the nuclear chain reaction would be harnessed for warfare became apparent fairly early. Europe was already heading towards conflict when Enrico Fermi did his groundbreaking work on nuclear reactions.

In 1923, Enrico Fermi realized that Einstein's equation $E=mc^2$ suggested that an enormous amount of energy could be released by destroying matter at the atomic level. He experimented with firing neutrons at uranium and other materials, and found that new elements seemed to be formed. These were isotopes, formed by blasting neutrons out of the atom. It was Leó Szilárd, in 1933, who suggested how the process of breaking atoms apart to release energy could be achieved. If a neutron were fired at high speed to blast an atom and release two neutrons in the collision, each of those neutrons could then interact with a further atom, releasing another two neutrons, and so on. This would set in motion a self-perpetuating reaction – a nuclear chain reaction – releasing large amounts of energy at every stage and a massive burst of energy overall. Szilárd's stunning realization of how to generate nuclear energy famously came to him as he waited at a red traffic light in London, where he was working to help refugee academics escape the Nazi regime.

After leaving Italy to pick up his Nobel Prize in 1938, Fermi went straight to the USA where he joined the Manhattan Project. There, he began the first-ever nuclear chain reaction at the reactor he had built under the bleachers of a sports field in Chicago. Not long after, the first – and only – atom bomb attacks took place on Hiroshima and Nagasaki.

FLASHPOINT FACT

The essay in physics that Enrico Fermi wrote as part of his entrance exam for university, the Scuole Normale Superiore in Pisa, was considered worthy of a student applying for a PhD place. He was 17 years old.

TIMELINE

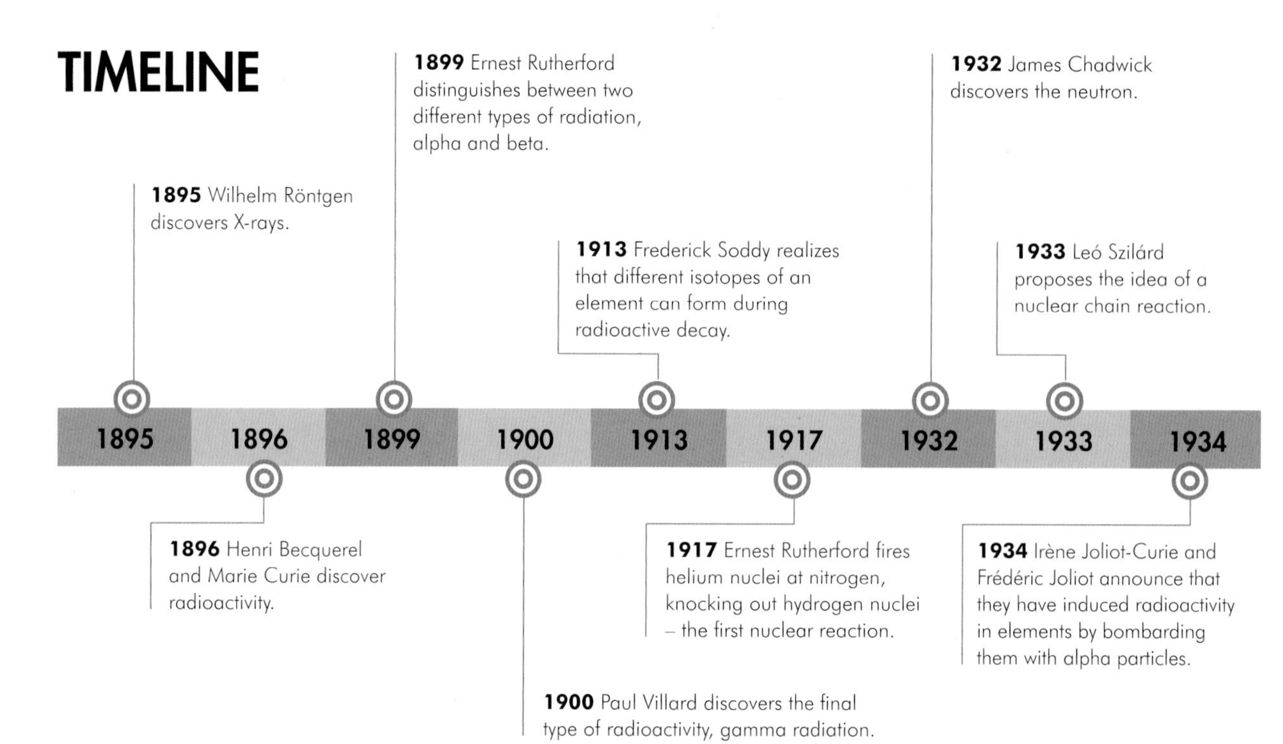

FLASHPOINTS

1896

RADIATION

Radiation is the natural emission of helium nuclei (alpha particles) or electrons (beta particles). It was first discovered by Henri Becquerel and his PhD student Marie Curie in 1896. Wilhelm Röntgen had discovered X-rays the previous year, and in 1900, Paul Villard discovered the shorter-wavelength X-rays that are the third type of radiation, gamma rays. In 1913, Frederick Soddy charted the existence of 40 apparently different elements involved in the radioactive decay of uranium to lead, although the periodic table has space for only 11. His conclusion was that there are variants of an element with different numbers of neutrons and protons, while the element retains essentially the same physical and chemical characteristics. These are radioisotopes.

1941

THE MANHATTAN PROJECT

The top-secret Manhattan Project, set up in 1941, was – it seemed – a race against time. It was instituted by the USA, UK and Canada to develop a nuclear weapon ahead of Nazi Germany producing one. Leó Szilárd and Eugene Wigner had written to President Roosevelt in 1939 warning him that the Nazis were working on nuclear potential and recommending that the US should get there first. Einstein was among the signatories. Some of the world's greatest scientists, including Einstein, Fermi and Niels Bohr, worked on the Manhattan Project. The first test of a nuclear weapon took place in New Mexico in July 1945; the following month, the US dropped atomic bombs on the Japanese cities of Hiroshima and Nagasaki.

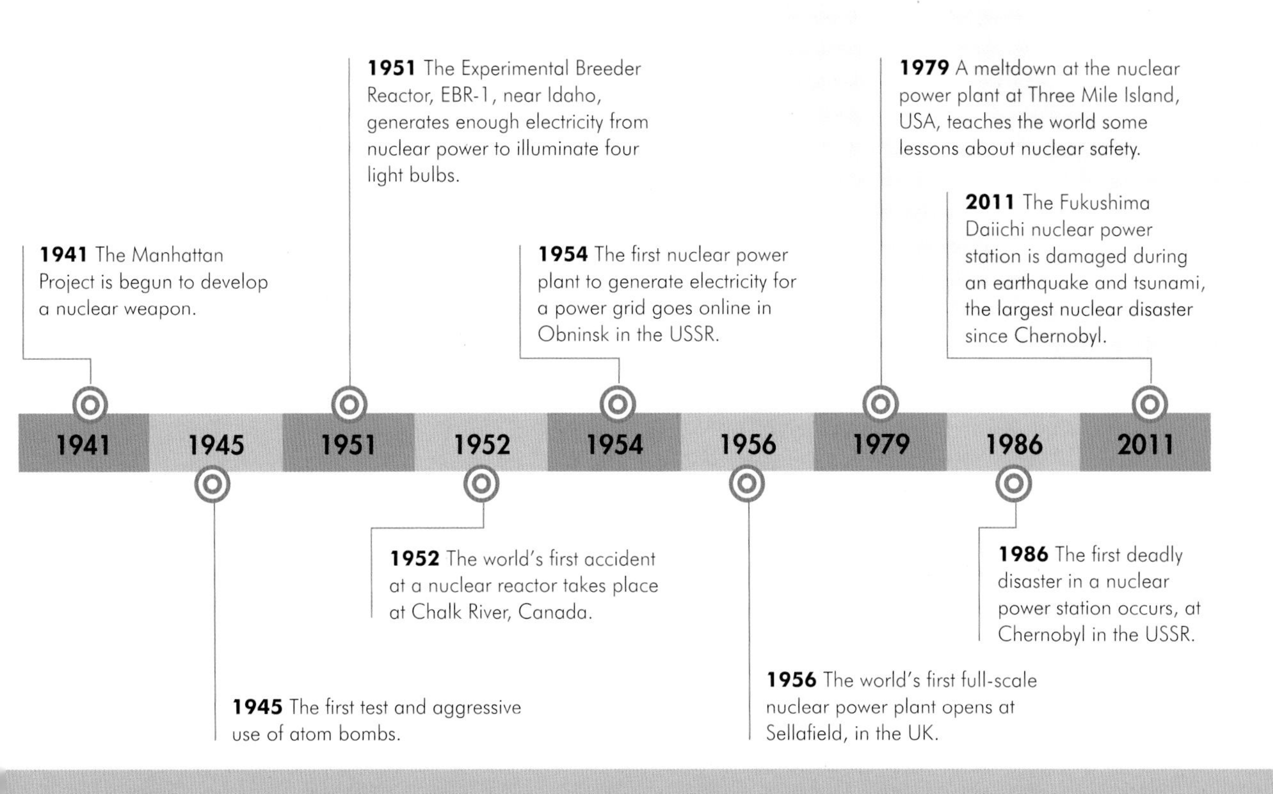

1948–54
FROM BOMBS TO BULBS

After the devastating effects of the atom bomb on Japan became known, many of the physicists who had been involved in the Manhattan Project had second thoughts. Some became peace activists. Work on atomic energy release found a new direction in developing nuclear power as a 'clean' alternative to burning coal and gas, and for powering ships and submarines. The test of this concept came in 1948, when nuclear power successfully generated enough electricity to light a bulb. Experimental nuclear power stations were built in the USSR, the USA and Great Britain during the 1950s. The first full-scale nuclear power station to be used to supply electricity for domestic use was Calder Hall-1 at Sellafield in Britain, opened in 1956. The first nuclear-powered submarine was USS *Nautilus*, launched in 1954.

Post 1979
MORE DISASTERS

The downside of nuclear power stations first became apparent with a lucky escape – the meltdown at a power station on Three Mile Island, USA, in 1979, which had no casualties – and then a disaster. A major accident in 1986 at the power station in Chernobyl, USSR, killed 31 people immediately but seriously affected the health of thousands more and has left tracts of land contaminated and unusable. The accidents crystallized opposition to nuclear power on grounds of safety, and led to the introduction of more stringent safety regulations. Catastrophic damage to the Fukushima Daiichi nuclear power station in Japan in 2011 reignited worries about the vulnerability of reactors.

The nuclear reactor in Chernobyl.

1942

BLASTING INTO SPACE

'It is difficult to say what is impossible, for the dream of yesterday is the hope of today and the reality of tomorrow.'

– Robert H. Goddard

The launch of a German V-2 rocket, 1942.

The launch of the first rocket that entered space – defined at the time as 80km (49 miles) above sea level – came in 1942 as part of the German development of rocketry for carrying bombs.

Wernher von Braun had hopes of making a rocket for space travel, and as a younger engineer joined an amateur rocketry club, *Verein für Raumschiffarht* (Society for Space Travel). But it was the 1930s and the Nazis were looking for new weapons they could develop. The VfR was awarded $400 to develop a rocket. Von Braun's first rocket, the A4, was renamed the V-2 and used to deliver bombs. As the A4 made the transition to a weapon, development moved out of Berlin onto an island on the Baltic coast of Germany, Usedom. Here, secrecy was more easily preserved and there was more space for testing. The first test crashed into the sea, the second exploded at an altitude of 11km (7 miles), and the third flew to 80km (50 miles) and landed on target 193km (120 miles) away.

After the war, the USA was keen to appropriate German scientific expertise and took in many scientists who might otherwise have gone to the USSR. Von Braun surrendered to the US with his 500-strong rocketry development team. To begin with, the impetus of rocketry development was still weaponry. But in 1958, NASA was formed in response to the Soviet launch of the satellite Sputnik, and interest turned towards space again – this time in a frenzied rush. Von Braun was invited to NASA in 1960 and became director of the Marshall Space Flight Center, where he directed the Saturn V launch and worked towards rockets later used in the moon landings.

The principle of the rocket, expelling exhaust gases at very high pressure from the rear to propel the rocket forward, relies on burning a large amount of fuel very quickly. A limiting factor is how much fuel can be packed into the rocket and – when it is in space – supplying oxygen to allow it to burn. Multi-stage rockets have extra fuel-carrying stages.

FLASHPOINT FACT

All three rocketry pioneers, Konstantin Tsiolkovsky, Robert Goddard and Hermann Oberth, were influenced by the science fiction story *From the Earth to the Moon*, written by Jules Verne and published in 1865.

TIMELINE

1903 | 1914 | 1920 | 1923 | 1924 | 1937

1903 Konstantin Tsiolkovsky publishes a book that includes the calculations to send a rocket into space.

1914 Robert Goddard registers two patents for rockets – one for a rocket using liquid fuel and one for a two- or three-stage rocket using solid fuel.

1920 Goddard's suggestion that a rocket might reach the moon is ridiculed in the American press.

1923 Hermann Oberth publishes a book about rocket flight into outer space.

1924 The Society for Studies of Interplanetary Travel is formed in the USSR, rooted in Tsiolkovsky's work.

1937 Oberth, von Braun and others form the Society for Space Travel in Germany, and go on to design the A-4 rocket.

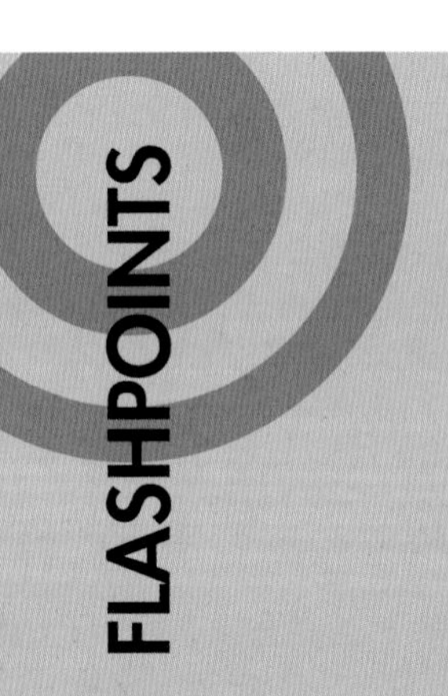

904AD
EARLY ROCKETS FOR FUN AND WAR

The principle of rocketry is very old. The Chinese were making rockets as fireworks and weapons hundreds of years ago; the use of fire arrows is first reported in 904AD during the siege of Yuzhang. A rocket works by explosively igniting fuel to create gases under high pressure that are pushed out of the back of the missile. This provides thrust, pushing the missile forwards. The fire arrow was propelled by a bag of gunpowder attached to the shaft. But the idea of a rocket powerful enough to go into space, or to propel a weapon a very long way, belongs firmly in the 20th century.

1903
REACHING FOR THE STARS

The first person to calculate that a rocket could be sent into space, and to write seriously about space travel, was the Russian scientist and philosopher Konstantin Tsiolkovsky in 1903. His work was little known in Europe and resulted in no practical developments. The American Robert Goddard launched a small, liquid-fuel powered rocket in 1926. It flew for only two-and-a-half seconds, reaching an altitude of just 12.5m (41ft) and landing 56m (184ft) away in a cabbage patch. Even so, it proved the concept of a rocket powered by burning liquid oxygen. He had said as early as 1920 that he could conceive of a rocket reaching the moon, though few others took this vision seriously.

Goddard (second from left) with his four-stage rocket.

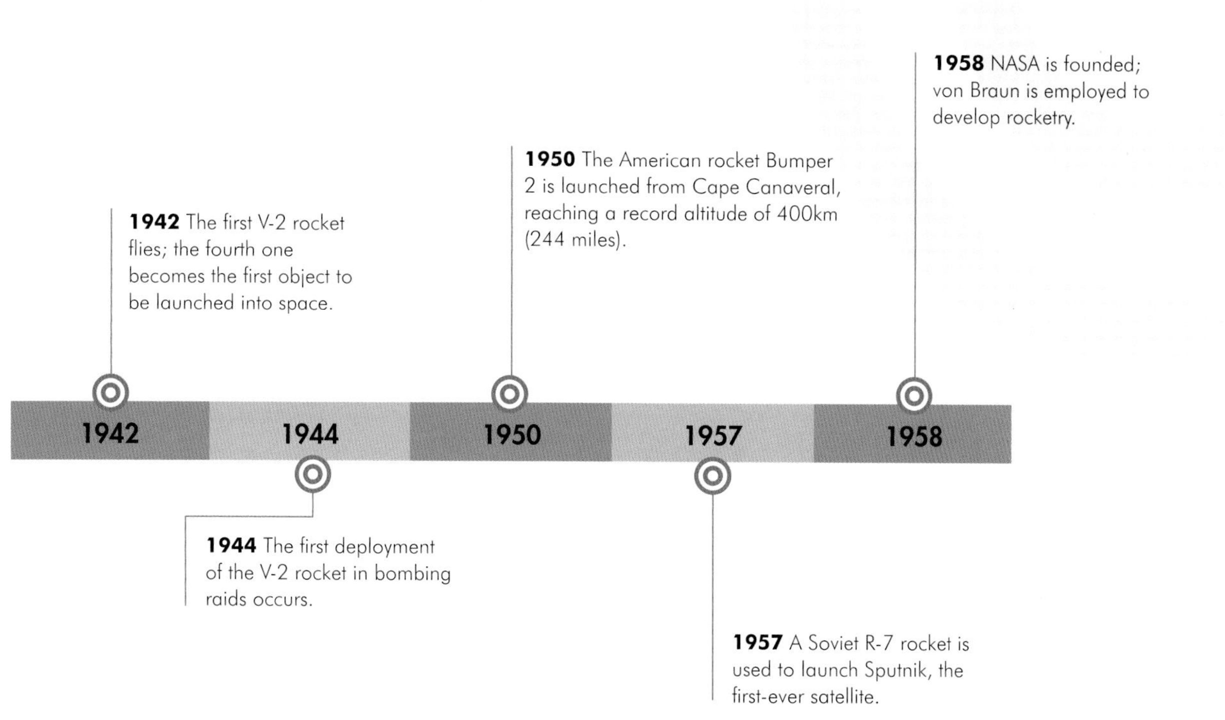

1923
ENTHUSIASTS LEAD THE WAY

The USSR's greatest proponent of space travel, Tsiolkovsky, was a mathematics teacher by day who wrote about and investigated rocketry in his spare time. In the west, lots of small groups of enthusiasts sprang up in Germany and elsewhere after Hermann Oberth published a book on space travel in 1923. There was no official funding for research into rocketry or space travel until it became clear that rockets could be used to carry bombs.

It was from this group of enthusiasts that von Braun and the A-4 emerged. When his rocket was perfected, it was not for his original purpose of space exploration, but as a weapon – Nazi funding was available for war but not for space.

1942
SHATTERING SILENCE

V-2 rockets travelled faster than the speed of sound, taking only five seconds from launch in the Netherlands to hit London. They arrived silently, the noise of them whooshing through the air being audible only after they had hit their target. In later days, when the rockets to come out of von Braun's department at NASA began to carry payloads into space, the rockets would stream silently ahead of their own sounds, travelling at speeds of nearly 10,000kph (6,214mph), or nine times the speed of sound.

A V-2 rocket ready for the launch pad.

1947

SHRINKING TECHNOLOGIES WITH THE TRANSISTOR

'About 60 million transistors were built this year just for you, with another 60 million for each of your friends, plus 60 million for every other man, woman and child on Earth. By 2010, the number should be around one billion transistors per person, per year.'

– Jim Turley, 2002

William Shockley, John Bardeen and Walter H. Brattain working at Bell Laboratories.

The tiny technologies we have today, such as smartwatches, smartphones and tablet computers, all owe their existence to the transistor, invented in 1947 by Bill Shockley, John Bardeen and Walter Brattain working for Bell Laboratories.

Before transistors, electronic equipment such as radios and televisions relied on large vacuum tubes, called diodes, to modulate and amplify electrical signals. The transistor replaced the vacuum across which the stream of electrons had to travel with a tiny gap in an electric circuit, and the filament-and-plate arrangement of cathode and anode with a tiny, fractured V-shape of metal. These innovations made the equipment needed much smaller, and were the first step in shrinking technology to become portable and eventually wearable.

Bardeen, Shockley and Brattain had been investigating the properties of semi-conductors in the 1930s, and after the Second World War looked to them again when seeking a replacement for the diodes in radios. They discovered that a germanium crystal with impurities made a better rectifier (allowing alternating current to pass in only one direction) than either a crystal or a tube, and also worked as an amplifier. As a semi-conductor, germanium neither carries electrons as well as a conductor such as metal, nor inhibits their movement as much as an insulator such as rubber. You could think of it as conducting electricity rather reluctantly. The impurities added make areas either better able to conduct electricity or less well able to do so. The two types can be carefully positioned to make the transistor act as a switch (stopping or starting current) or as an amplifier (boosting a small signal). Modern transistors have been further miniaturized so that billions can fit on an integrated circuit board the size of a fingernail.

FLASHPOINT FACT

The Apollo Guidance Control had around 10,000 transistors; a modern smartphone has about two billion, and a personal computer about two trillion.

TIMELINE

1904 John Ambrose Fleming invents the vacuum tube.

1931 Alan Wilson publishes *The Theory of Electronic Semi-Conductors*, which uses quantum theory to explain semiconductor properties.

1936 The printed circuit is invented by Paul Eisler.

1943 The USA starts to use printed circuits in military equipment in the Second World War.

1947 Shockley, Bardeen and Brattain working at Bell Laboratories develop the first transistor.

1948 Printed circuit technology is released for commercial use.

1952 Radios and hearing aids including transistors go on sale; consumers are willing to pay high prices.

1953 The first high-frequency (60 MHz) transistor is produced by Philco.

FLASHPOINTS

1904
NOTHING TO SEE

The earliest radios depended on vacuum tubes as a means of amplifying or modifying an electrical signal. The design is simple: a tube containing no gas or other air has a filament that heats up when an electric current is passed through it and acts as a cathode, emitting electrons. These are attracted across the vacuum to a metal plate or film anode. By putting a control grid inside the tube, the stream of electrons can be amplified or modified. Vacuum tubes made possible radio, television and early computers. Although they were effective, they were both heavy and fragile. They were also inefficient in terms or energy use, and took time to warm up.

A radio valve.

1947
WHAT IT DOES

A transistor can operate as a switch or as an amplifier. It works by using a small amount of electricity (current) to control a large supply of electricity by operating a 'gate'. There are three basic components: a base (which is the gate), a collector (which is the larger supply of electricity) and an emitter (the outlet for the supply). Either a small amount of current through the gate can regulate the amount of current passing through the main supply (acting as an amplifier), or it can be used in a binary form: if enough current passes, the gate opens, otherwise it is closed. In this mode, it operates as a switch, turning the current on or off.

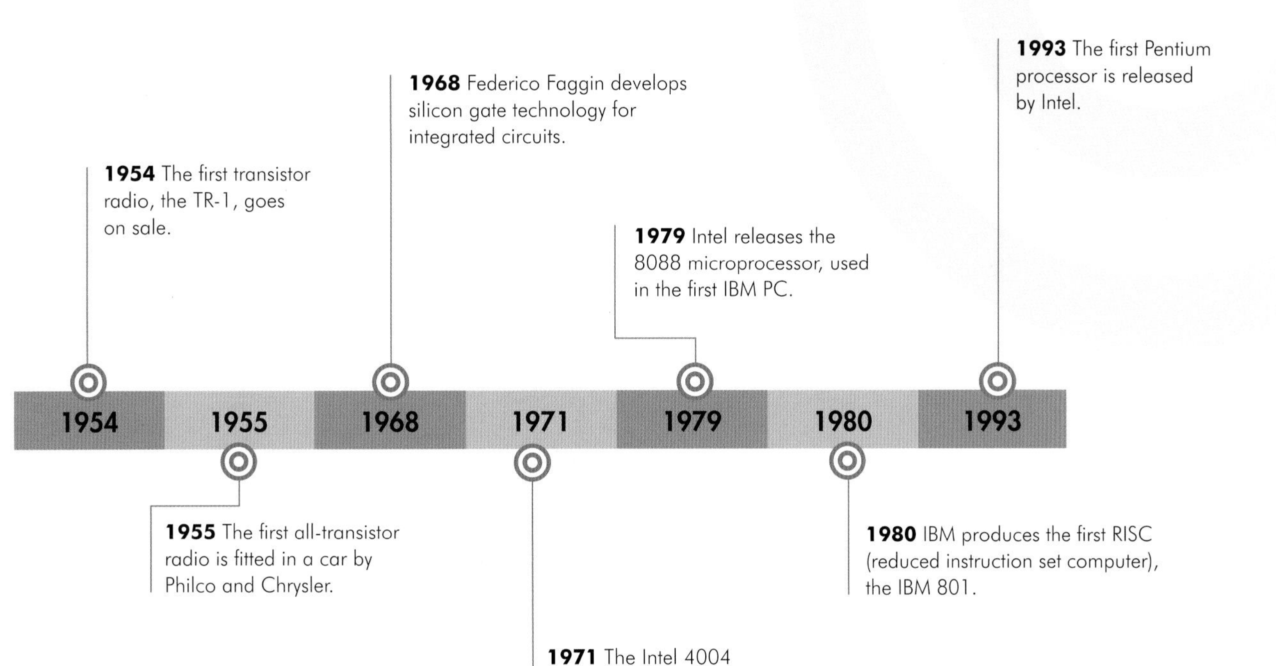

1948
PRINTED COMPONENTS

Although billions of individual transistors are still produced today, even more are incorporated into integrated circuits. These produce entire electrical circuits by printing (or etching) onto semiconductive silicon wafers. The electric current flows along lines printed in metal that make up the circuit, sometimes only a few tens of nanometres wide. The number of transistors in a circuit tends to determine its processing power. Modern integrated circuits can fit 250 billion transistors into an area the size of a fingernail. Your smartphone has more than a billion transistors, and they can now be spaced only 20 nanometres apart.

1954
SMALLER IS (USUALLY) BETTER

Transistors offered an improvement over vacuum tubes in many regards. Other than the most obvious advantage of them being much smaller and lighter, they are also more robust, last longer and require little enough power to be run from a small, portable battery. Individual transistors, even when newly developed, were much smaller than vacuum tubes. When printed onto circuit boards, they can be microscopically small. This means that thousands – even millions – can be crammed into the smallest of devices. The transistor radio, invented in 1954, was a world-changing technology, bringing portable music and news to everyone. The first model sold 100,000 units in the first year.

A printed circuit board crams many transistors into a tiny space.

1953

THE SPIRAL OF LIFE: DNA

'It has not escaped our notice that the specific pairing we have postulated immediately suggests a possible copying mechanism for the genetic material [DNA].'

– Francis Crick and James Watson, 1953

Rosalind Franklin, 1955.

DNA is the complex molecule that codes genetic material. In its structure are all the instructions for making an individual organism, whether it's a bacterium, a human being or an extinct mammoth.

That knowledge was slow in arriving, taking more than a century to emerge as scientists approached the matter of inheritance from two angles: from observing how characteristics are passed from parent to progeny, and looking at the mechanisms and biochemistry of the cells involved in reproduction. Even after it was recognized that DNA is the molecule that makes up genes, its structure remained obscure until revealed by Francis Crick, James Watson and Rosalind Franklin in 1953.

Just knowing the chemicals involved in a large molecule is only a very small first step; the arrangement of hundreds or thousands of atoms is what gives a compound its properties. Rosalind Franklin's special talent was X-ray crystallography – using X-rays bounced off molecules to reveal their structure. Her supervisor Maurice Wilkins showed her photograph of a DNA molecule to Crick and Watson without her permission. From it, they worked out the crucial detail that DNA is two strands twisted together into a helix, with bridges holding the strands equidistant along the length, like the rungs of a ladder. The bridges are made of bases – repeated molecular units of a particular configuration. There are four bases, which occur in paired proportions. Every adenine must be paired with a thymine, and every cytosine must be paired with a guanine. This means that the process for making DNA in a cell is straightforward: the double helix can unzip into two strands and each strand can rebuild its partner strand because the bases are a code to its construction – where there is a cytosine, a guanine is needed opposite it, and so on.

FLASHPOINT FACT

After working out the crucial details of the structure, Crick and Watson crossed the road from the Cavendish Laboratory in Cambridge and celebrated in the Eagle pub.

TIMELINE

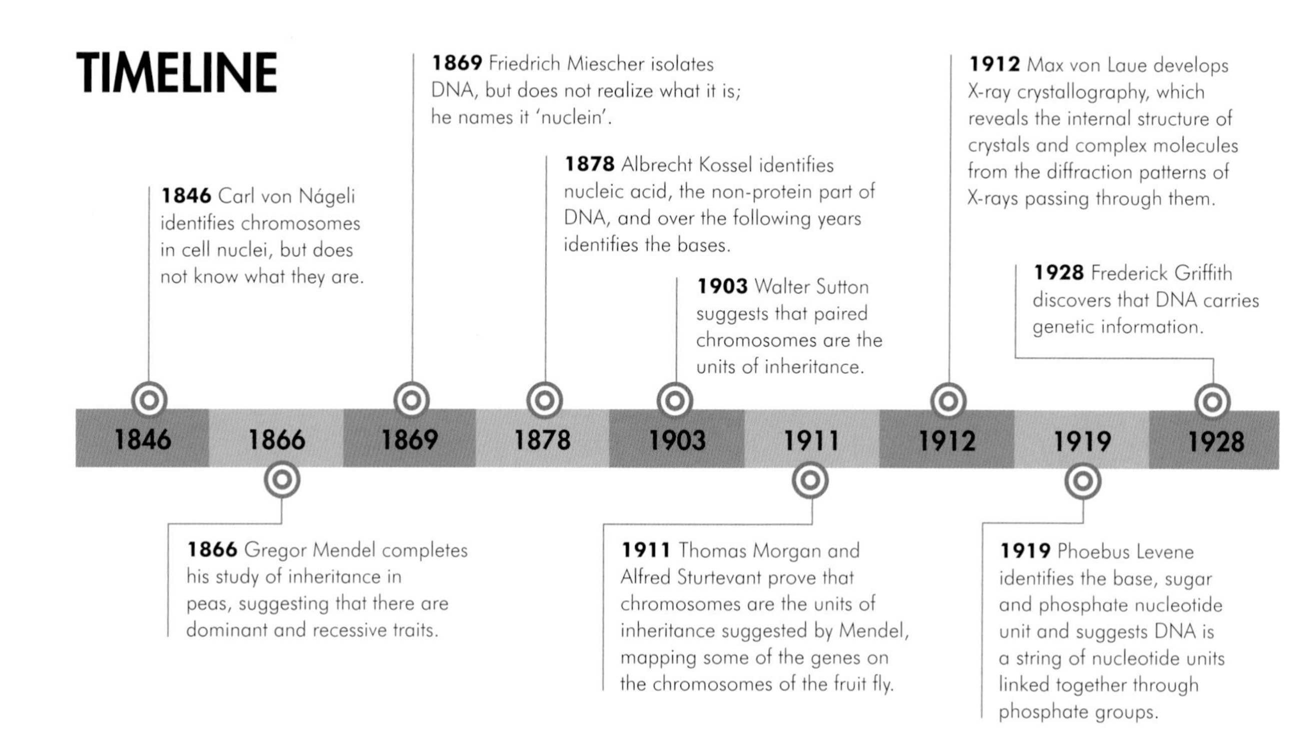

FLASHPOINTS

1869
PUS AND PEAS

The first person to isolate DNA was Friedrich Miescher who, in 1869, gathered a large number of pus-soaked bandages from a hospital and extracted the cell nuclei of white blood cells. He found a substance that he called 'nuclein' in all the cells he examined, but did not know its function. At around the same time, the Moravian monk Gregor Mendel was experimenting with breeding peas and gaining an insight into how heredity worked. He was able to explain something that farmers had discovered long before about inherited characteristics in plants and animals. He deduced that some characteristics are 'dominant' and some 'recessive' and that if progeny are to manifest a recessive characteristic, they must inherit it from both parents.

Gregor Mendel, who gained posthumous fame as the founder of the modern science of genetics.

1911
CHROMOSOMAL PAIRS

Mendel's work went largely unnoticed until it was rediscovered in 1900. Still it had no immediate impact. But then Walter Sutton, studying meiosis (cell division in the production of sperm and egg cells) noticed that chromosomes seemed to occur in pairs, but that the pairs split up in meiosis. He suggested that the paired chromosomes were the mechanism of inheritance that Mendel had postulated. The mechanism worked with Mendel's findings; progeny could inherit two recessive markers from parents that both manifested the dominant marker on their physiology. His theory was proved correct in 1911 by Thomas Morgan and Alfred Sturtevant. Working with fruit flies (*Drosophila*), they showed how genes can be linked if they occur on the same chromosome.

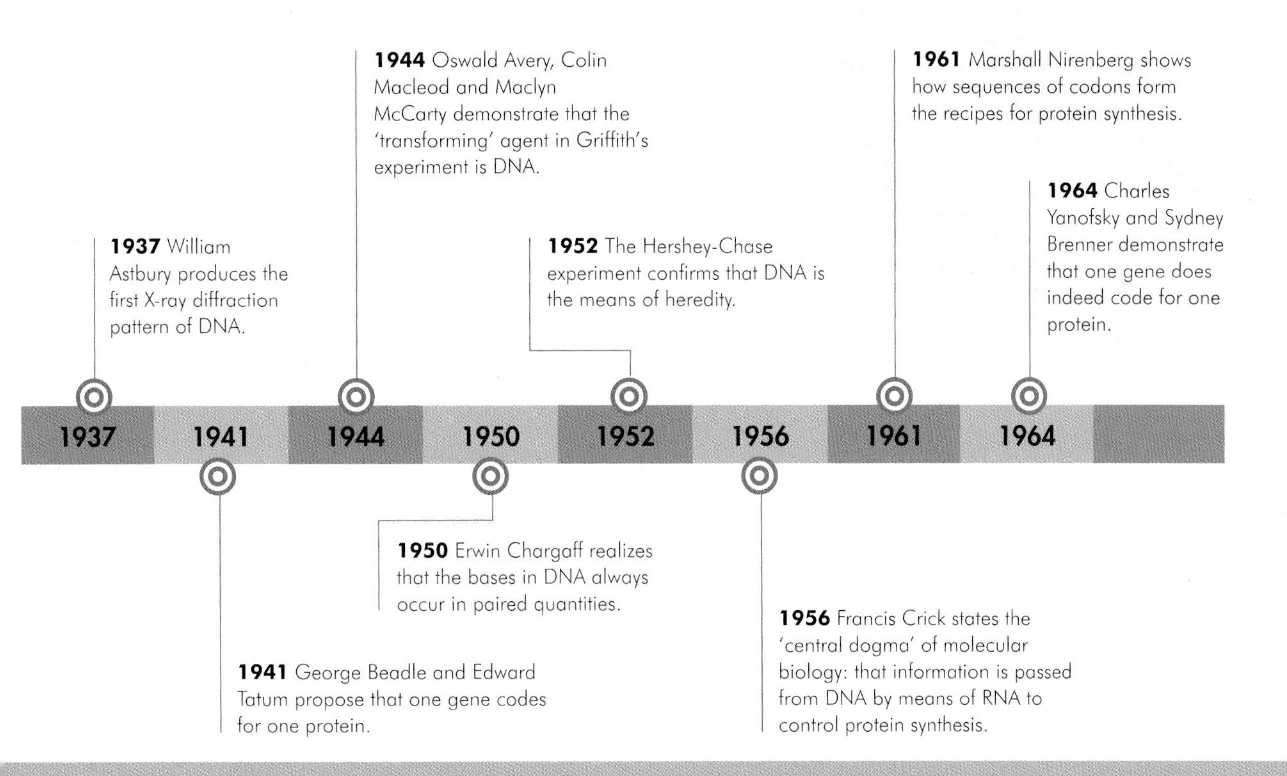

1944
DNA'S THE STUFF

In 1944, Oswald Avery, Colin Macleod and Maclyn McCarty were experimenting with two strains of pneumonia, one virulent and one not. They had discovered that an extract from dead virulent bacteria could transform live 'safe' bacteria into the deadly form. They found that if they treated the extract with an enzyme to break down protein, the transformation still worked, but if they treated it with an enzyme that breaks down DNA, it no longer worked. This proved that it was DNA that carried the information and was being absorbed into and modifying the previous 'safe' bacteria. It was the first demonstration that DNA is, in fact, the material of inheritance.

1961
DNA MAKES RNA MAKES PROTEINS

How does the coding of thousands of sets of base pairs on a molecular strand translate into growing blue eyes or the right number of fingers? This is all down to making the right proteins, in the right place, at the right time. Francis Crick stated this 'central dogma' of molecular biology in 1956, explaining how the code of DNA is read and actioned by RNA. Information, he said, goes from DNA to RNA to proteins – it's a one-way flow. Marshall Nirenberg discovered in 1961 that the sequence of bases is divided into triplets, called codons, each coding for a specific amino acid. Amino acids are the building blocks of proteins. By putting the amino acids together following the instructions in DNA, the right protein is made each time.

1954

NEW FOR OLD: ORGAN TRANSPLANTS

'It is infinitely better to transplant a heart than to bury it to be devoured by worms.'

– Christiaan Barnard, the first surgeon to perform a heart transplant

Ronald (left) and Richard Herrick.

For organ transplants to work, we now know, the recipient's body must be receptive to the donated organ. This knowledge comes from the first successful kidney transplant, carried out in 1954. Ronald Herrick donated a kidney to his brother Richard. The secret to the success was in their genetic makeup: Ronald and Richard were identical twins, so Richard's immune system accepted Ronald's kidney rather than rejecting it. Identical twins share the same DNA, so to the immune system, Richard's and Ronald's kidneys were indistinguishable. Richard lived a further eight years after the operation.

Surgeons operating to transplant his twin's kidney into Richard Herrick.

Prior to the Herricks' experience, the French surgeon Jean Hamburger carried out a transplant in 1952, using a kidney from the patient's mother. The organ was rejected after three weeks, the recipient's body treating it as foreign matter and setting out to destroy it. After the success of the Herrick transplant, scientists turned to finding a method of calming the immune system. The result was anti-rejection drugs, the most successful of which, cyclosporine, was developed in 1972. The difficulty with anti-rejection drugs, though, is that because they suppress the immune system, the patient is put in danger of contracting other diseases and being unable to fight them off. Transplant patients must take anti-rejection drugs for the rest of their lives. However, in spite of the considerable and ongoing difficulties, there is no doubt that organ transplants have provided hope to millions who would otherwise have been resigned to death.

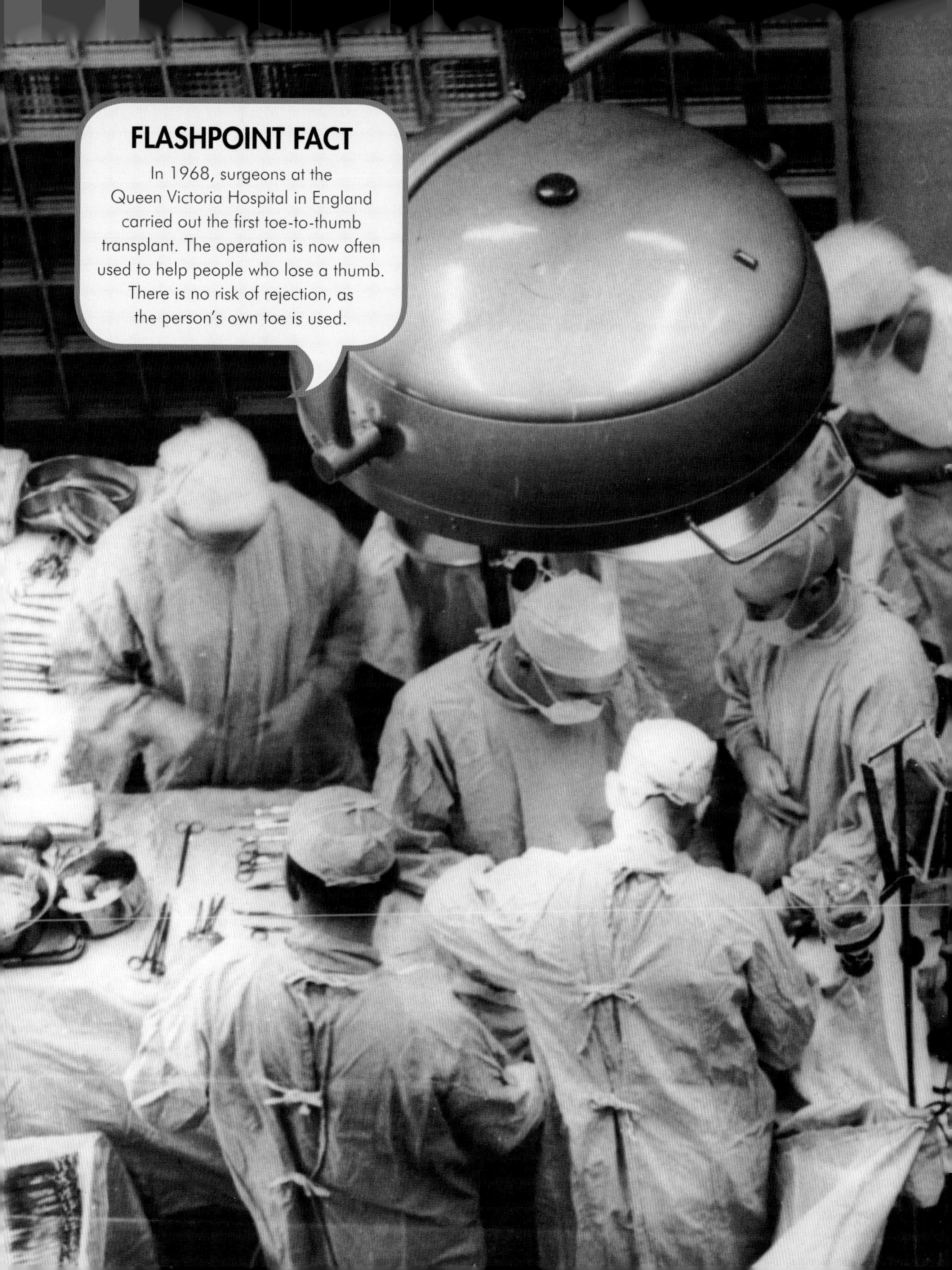

FLASHPOINT FACT

In 1968, surgeons at the Queen Victoria Hospital in England carried out the first toe-to-thumb transplant. The operation is now often used to help people who lose a thumb. There is no risk of rejection, as the person's own toe is used.

TIMELINE

1668 | 1881 | 1901 | 1905 | 1908 | 1936 | 1940s | 1954

1668 A collection of case histories compiled by Dutch physician Job van Meekeren claims that a Russian surgeon has managed the first dog to human skin graft.

1881 A skin graft from a dead body helps heal a man burned by leaning against a metal door as it was struck by lightning.

1901 Karl Landsteiner identifies different blood groups.

1905 Dr Eduard Zirm performs a successful cornea transplant, restoring the sight of a farm labourer blinded in an accident.

1908 Alexis Carrell perfects the fine stitching needed for joining blood vessels in transplant surgery.

1936 Yuri Voronoy in the USSR performs a human-to-human kidney transplant, but the recipient dies after two days.

1940s Zoologist Peter Medawar experiments with skin grafts with animals and discovers that an immune response leads to rejection.

1954 The first successful kidney transplant takes place.

FLASHPOINTS

1668
SKIN AND BONES

There are a few, probably apocryphal, accounts of early transplants, but the first that is likely to be true is a bone transplant from a dog to patch the injured skull of a Russian soldier in 1668. Assuming the bone was thoroughly cleaned of all dog tissue, this would be no less likely to succeed than patching the skull with metal. Skin grafts using the recipient's own skin have been used since the ninth century in India to rebuild noses and treat burns and wounds. This ran into no problems of rejection since the skin (still attached by a flap to its former mooring place) was the patient's own.

A diagram explaining how to carry out nose reconstruction, 1820.

1909
NOT QUITE THERE

The first attempts at organ transplant were less successful. In 1909, a French surgeon tried to save a child with kidney failure by inserting slices of rabbit kidney, but the child died after a few days. Similar attempts with kidneys from other animals were equally unsuccessful. During this period, though, Alexis Carrell was pioneering techniques in suturing at a sufficiently small scale to connect blood vessels and nerves, which would be essential in later transplants. Landsteiner's discovery of blood groups removed another hurdle, and the development of antibiotics offered protection against post-operative infection. But attempts at organ transplant continued to fail until the experience of the Herricks highlighted the role of the immune system in rejection.

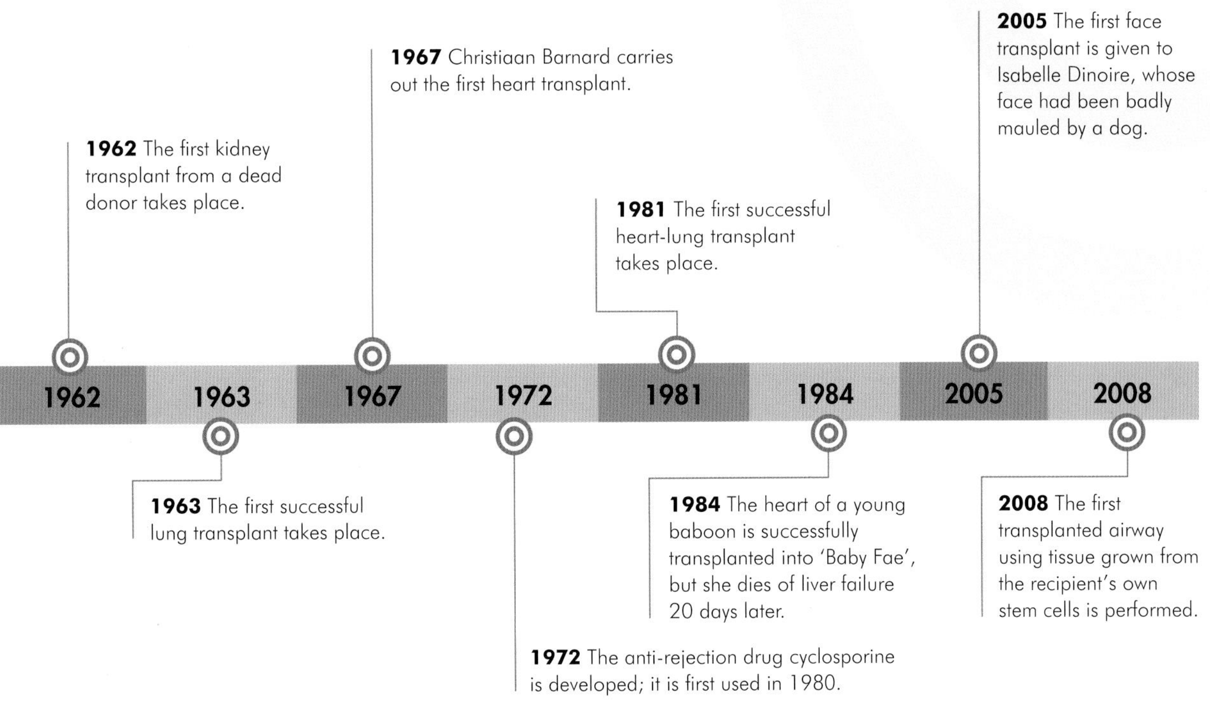

1967

HEART TO HEART

The Herrick kidney transplant was the first success, but the first heart transplant is the most famous transplant story of the 20th century. In 1967, South African surgeon Christiaan Barnard transplanted the heart of Denise Darvall, who had died in a traffic accident, into 54-year-old Louis Washkansky. The heart worked. Unfortunately, after 18 days Washkansky died of pneumonia that his suppressed immune system could not combat successfully. The heart had continued to function, though: the problem to sort out was to do with the anti-rejection medication. Once cyclosporine became available in the 1980s, more and more successful transplants of different and multiple organs followed.

2008

MADE TO MEASURE

The limiting factor for transplants became the supply of organs. In 2008, doctors tried to combine the two fields of organ transplant and regenerative medicine, in the world's first transplant of a regenerated airway. The scaffold for the airway came from a donor, but bone marrow cells from the patient were used to seed the scaffold. This meant that the tissue grown on the scaffold was the patient's own and so anti-rejection drugs were not needed. This was the first organ transplant that did not need anti-rejection drugs and was also not between identical twins. The same technique, but using a synthetic scaffold, was successful in 2011. No anti-rejection drugs are needed with a synthetic scaffold and there is no need to wait for a donor.

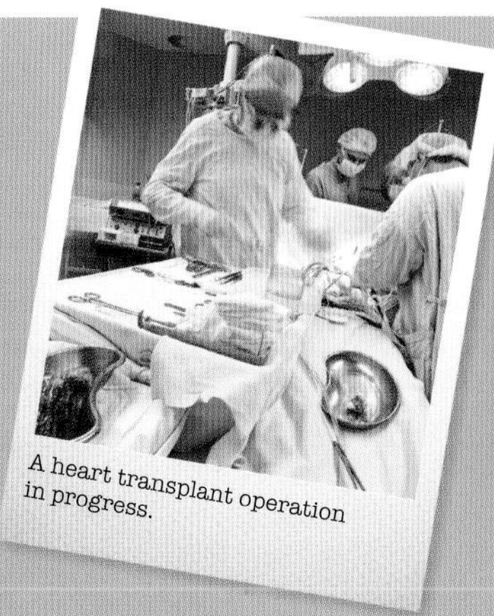
A heart transplant operation in progress.

1955

THE VACCINE THAT SAVED A NATION'S CHILDREN

'I have studied the effects of our new lots of polio vaccine...and...shall give it to my wife and two children as well as to our neighbours and their children.'

– Albert Sabin, 1957

Children and parents wait in line for polio vaccinations on the first day that the vaccine was made available to children in the USA.

Polio is a disease that attacks the nervous system, frequently leaving victims paralyzed and with wasted limbs. It became increasingly common in the USA during the early decades of the 20th century, leading to annual fear as epidemics often raged during the summer months. These became worse and worse as the century went on.

The quest for a vaccine began in earnest in the 1930s, after the successful development of a vaccine against diphtheria in 1923. The first trials, in 1935, were disastrous, leading to many test subjects developing polio. Some died, many were paralyzed and some suffered an allergic reaction to the vaccine.

Two researchers working independently spearheaded the hunt for a vaccine: Jonas Salk and Albert Sabin. Salk produced his vaccine first and began trials of it early, spurred to urgent action by the worst summer epidemic in the USA's history in 1952. Large-scale trials followed in 1954, funded by the March of Dimes and involving 1.3 million children. The vaccine proved effective and was licenced for use in 1955, to the jubilation of the American public. Salk's vaccine was made from polio virus that had been killed. Just a few weeks after mass vaccination began, some batches of incorrectly prepared vaccine caused cases of polio, resulting in 11 deaths and many cases of paralysis. Vaccination was suspended for a while, and the public lost confidence in it. Sabin continued to develop his oral vaccine, testing it extensively in the USSR in 1959. It eventually overtook Salk's, being easier to administer, cheaper and more effective.

Though polio was the most feared disease of the 20th century, the defeat of other deadly diseases including diphtheria and pertussis (whooping cough) through vaccination were equally important triumphs of medical science.

FLASHPOINT FACT

Up to 95 per cent of people who contract polio show no symptoms – but they can still pass the disease on to others, making it very easy for it to spread.

TIMELINE

1796 Edward Jenner develops the first vaccine, to protect against smallpox.

1892 Antitoxin for diphtheria is developed (given after the disease is contracted).

1894 The first epidemic of polio in the USA.

1914 William H. Park develops a mix of antitoxin and toxin as a vaccine against diphtheria.

1923 A better diphtheria vaccine, diphtheria toxoid, is developed.

1935 Early trials of a polio vaccine are disastrous, with some patients dying or being paralyzed.

1936 Max Theiler develops a vaccine against yellow fever.

1938 Entertainer Eddie Cantor initiates the March of Dimes to raise money for research into polio.

1942 The DTP vaccine is introduced, a combined vaccine for diphtheria, tetanus and pertussis (whooping cough).

1952 A massive polio epidemic in the USA leaves 21,000 people paralyzed.

FLASHPOINTS

1879
MAKING VACCINES

Pasteur developed a vaccine for chicken cholera in 1879 and for the rabies virus in 1885, successfully treating a young boy after he had been bitten by a rabid dog. The principle of making vaccines has remained the same since his time: to isolate the pathogen and then make a form of it that has either been killed (usually through treatment with heat or chemicals) or has been attenuated – weakened so that it can't cause the full disease. The part of the pathogen that causes illness is separated from the part that prompts an immune response. Supplying only the latter prompts the body to make antibodies, but presents no risk of illness.

1881
FINDING MICROBES

The idea that disease is caused by germs was originally only one of several theories. It was championed by Louis Pasteur and Robert Koch in the mid-19th century, and supported by their discoveries of bacteria that were responsible for particular diseases. Pasteur first discovered that microbes are responsible for fermentation and for food spoiling, and soon after discovered that they can be pathogens – things that cause disease. In 1881, he discovered the bacterium that causes pneumonia and meningitis; Robert Koch discovered that responsible for tuberculosis in 1882, and that for cholera in 1884; Edwin Klebs found the diphtheria bacterium in 1883. Pasteur also isolated the virus that causes rabies, but it was too small to see with contemporary microscopes.

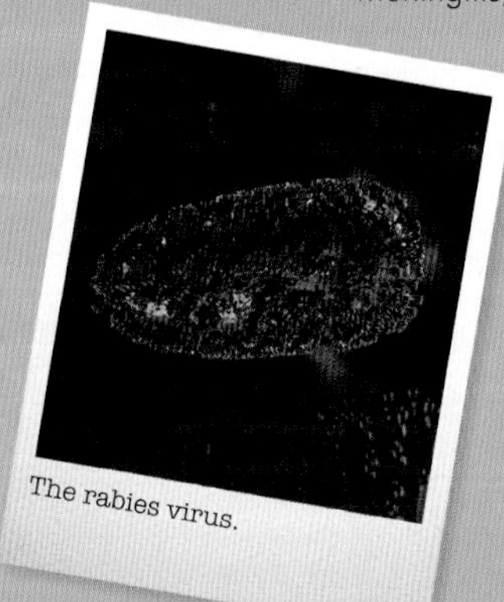
The rabies virus.

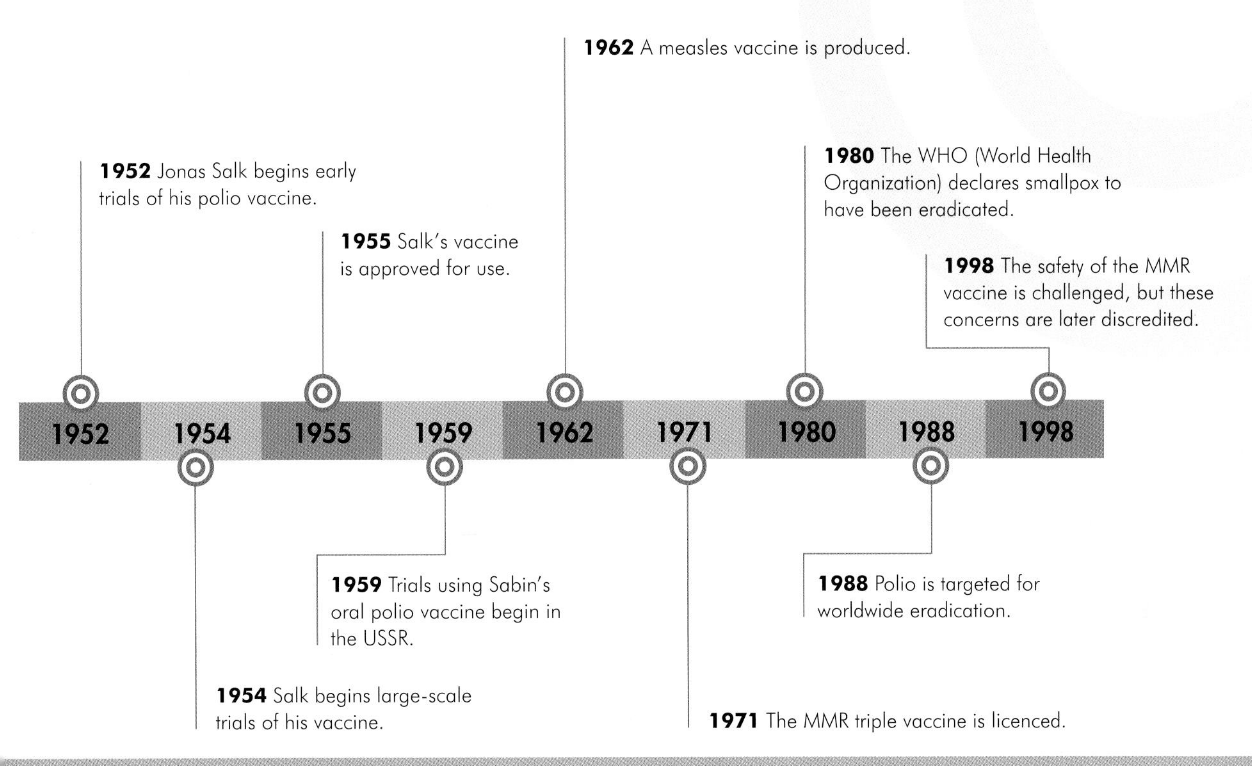

1914
THE PROMISE OF VACCINES

In the early 20th century, the search for vaccines to protect against some of the other terrible diseases that ravaged the population began. Among the first to be tackled was diphtheria, which can cause suffocation when membranes grow, blocking the throat. The first vaccine, developed in 1914, mixed the antitoxin given to affected patients with the diphtheria toxin itself to prompt immunity. A better vaccine was produced in 1923 and the terrible death rate of children from diphtheria fell dramatically. A vaccine for yellow fever followed in 1936. The triple vaccine for diphtheria, tetanus and pertussis (whooping cough) was introduced in 1942.

1971
MMR – GOOD WORK UNDONE

The triple vaccine against measles, mumps and rubella, MMR, was introduced in 1971. In 1978, measles was targeted for elimination and by 1981 cases had dropped considerably. But then in 1998, Andrew Wakefield threw a spanner in the works with a fraudulent claim that the MMR vaccine could lead to autism in vaccinated children. Vaccination rates dropped to below the level needed to sustain herd immunity, allowing the diseases to spread again. Although most of Wakefield's collaborators renounced their results, all subsequent research has found no link and Wakefield has been banned from practicing medicine on the grounds of misconduct, some people still believe the false research. Vaccination rates have not fully recovered and measles is back.

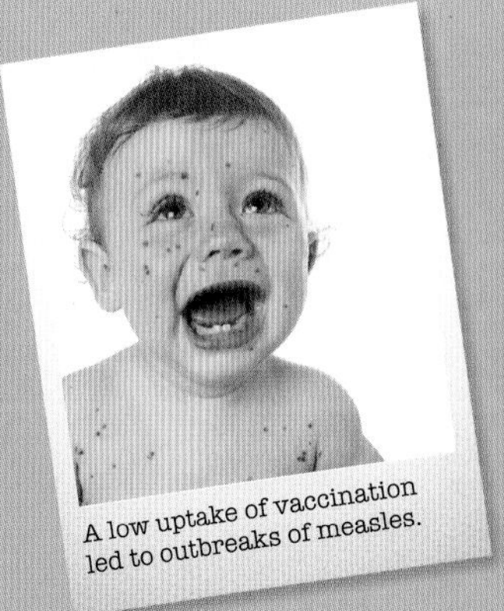
A low uptake of vaccination led to outbreaks of measles.

1957

SPUTNIK: FIRST STRIKE IN THE SPACE RACE

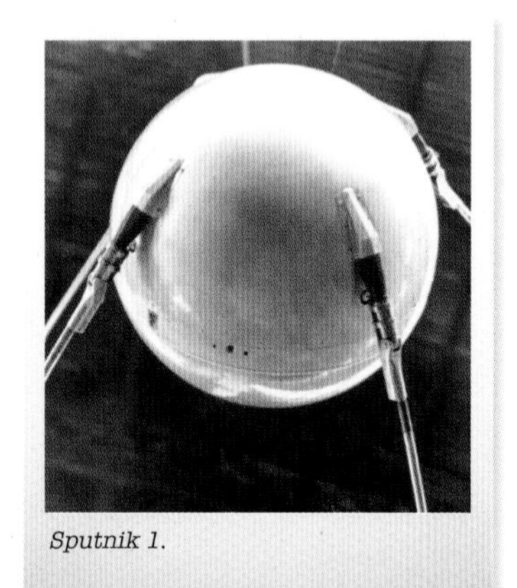
Sputnik 1.

The two halves of the spherical *Sputnik 1* separated to show the equipment inside.

'The new socialist society turns even the most daring of man's dreams into a reality.'

– TASS statement, 4 October 1957

The Space Race began when the Americans weren't looking; the first lap was won by the USSR on 4 October 1957 with the surprise launch of Sputnik, the first satellite, into orbit around the Earth. In fact, both superpowers had been aiming to launch a satellite since the mid-1950s, but the American effort was still several months away from launch. Their satellite was also much smaller, at around 7kg (15lb), than Sputnik's 83.5kg (184lb). Sputnik orbited at 900km (558 miles) above Earth, taking around 96 minutes to complete its elliptical orbit at a speed of almost 30,000kph (18,640mph).

Sputnik was designed to collect data on the upper atmosphere. Despite the entirely benign, scientific target (in this instance), many people in the USA were alarmed by the apparent supremacy of the USSR and feared that they could use a satellite to launch a nuclear weapon – it was the middle of the Cold War. Fear and humiliation spurred the US to urgent action, with the rapid formation of NASA and launch of their own satellite – though not the satellite that had been scheduled for launch. That now looked rather paltry beside Sputnik. The new satellite, *Explorer*, was hurriedly put together, and a rocket adapted to carry it in just 84 days. *Explorer* remained in orbit until 1970, re-entering the atmosphere over the Pacific Ocean on March 31 after more than 58,000 orbits. An identically constructed flight backup of *Explorer* is on display in the Smithsonian Institution's National Air and Space Museum in Washington, DC.

FLASHPOINT FACT

The orbit of *Sputnik* degraded and it broke up in the Earth's atmosphere during its descent on 4 January 1958.

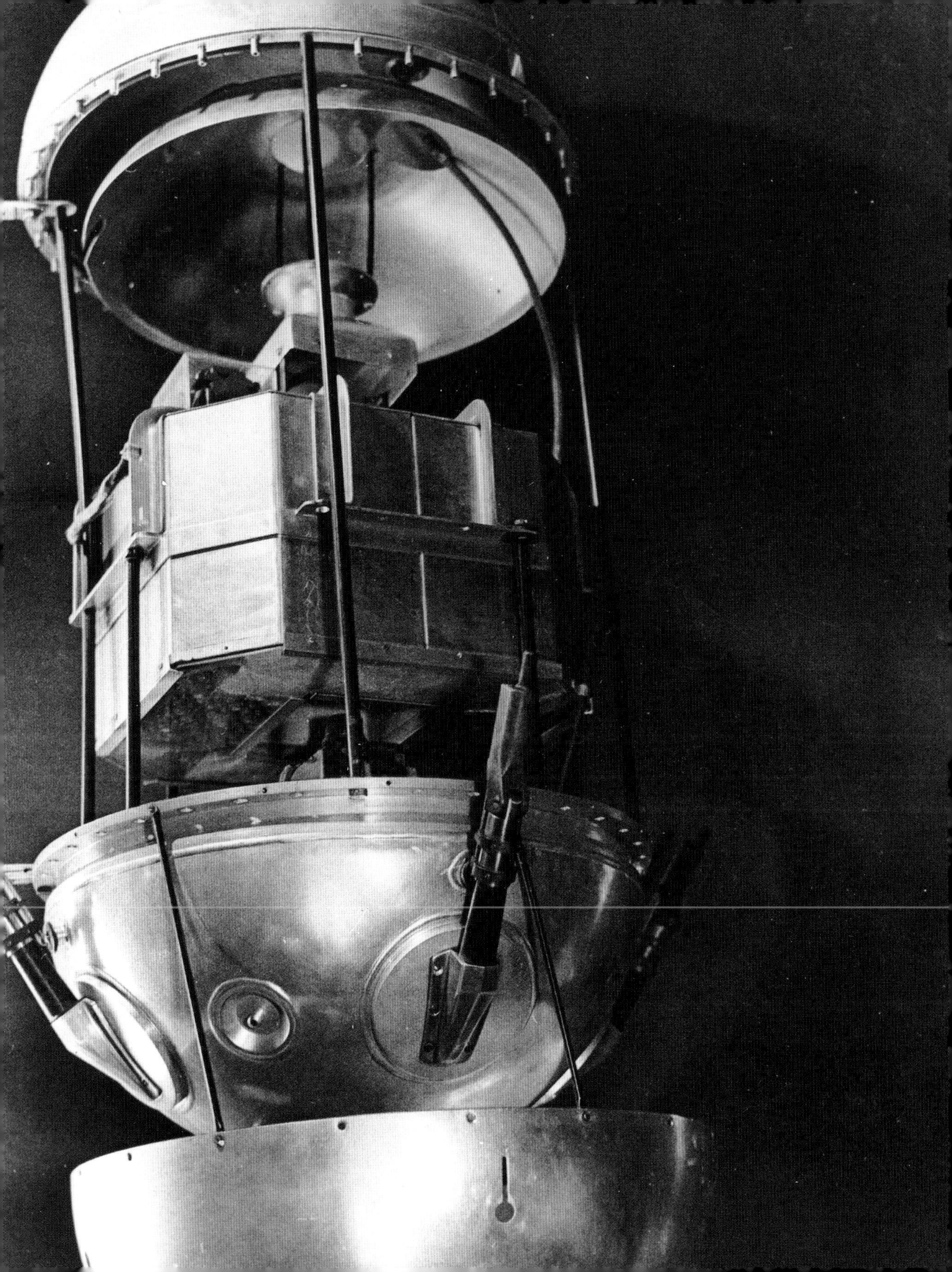

TIMELINE

1951 The USSR sends two dogs into space on a suborbital flight.

1952 The International Council of Scientific Unions names 1957–8 International Geophysical Year.

1954 The International Council of Scientific Unions calls for satellites to be launched in 1957 to map the surface of the Earth.

1957 *Sputnik*, the first artificial satellite, is launched by the USSR.

1957 *Sputnik* 2, carrying the dog Laika, is launched.

1958 NASA is founded to spearhead the US's space mission; the *Explorer* satellite is launched.

1960 The USA launches the first communications satellite, *Echo 1*.

FLASHPOINTS

1958
US TRAILING BEHIND

The first US satellite, *Explorer*, was launched a few months after *Sputnik*, on 31 January 1958. By then, the USSR had already launched a second satellite, *Sputnik 2*, carrying the dog Laika into space. The Soviet space programme remained well ahead of NASA in the space race through the 1950s and the early 1960s. The USSR achieved the first man in space, the first woman in space, the first space walk, the first spacecraft to impact the moon, the first craft to impact Venus, the first craft to orbit the moon and the first craft to achieve a soft landing on the moon. It was the USA, though, who achieved the ultimate goal of landing astronauts on the moon in 1969.

1960
SATELLITE BONANZA

Sputnik was designed to gather data about the upper atmosphere, and collecting atmospheric and weather-related data remains a task carried out by satellites. The first communications satellite, launched by the USA in 1960, was a giant silvered balloon, *Echo 1*, sometimes scathingly referred to as a satelloon. It provided a reflective surface from which radio signals could be bounced. Far from elegant, this solution had the disadvantage that there was rather random scattering, and it was a pretty hit-and-miss affair – but it set the telecommunications ball rolling (through space) and telecoms is now a major function of satellites.

A static inflation test of the *Echo I* communications balloon satellite.

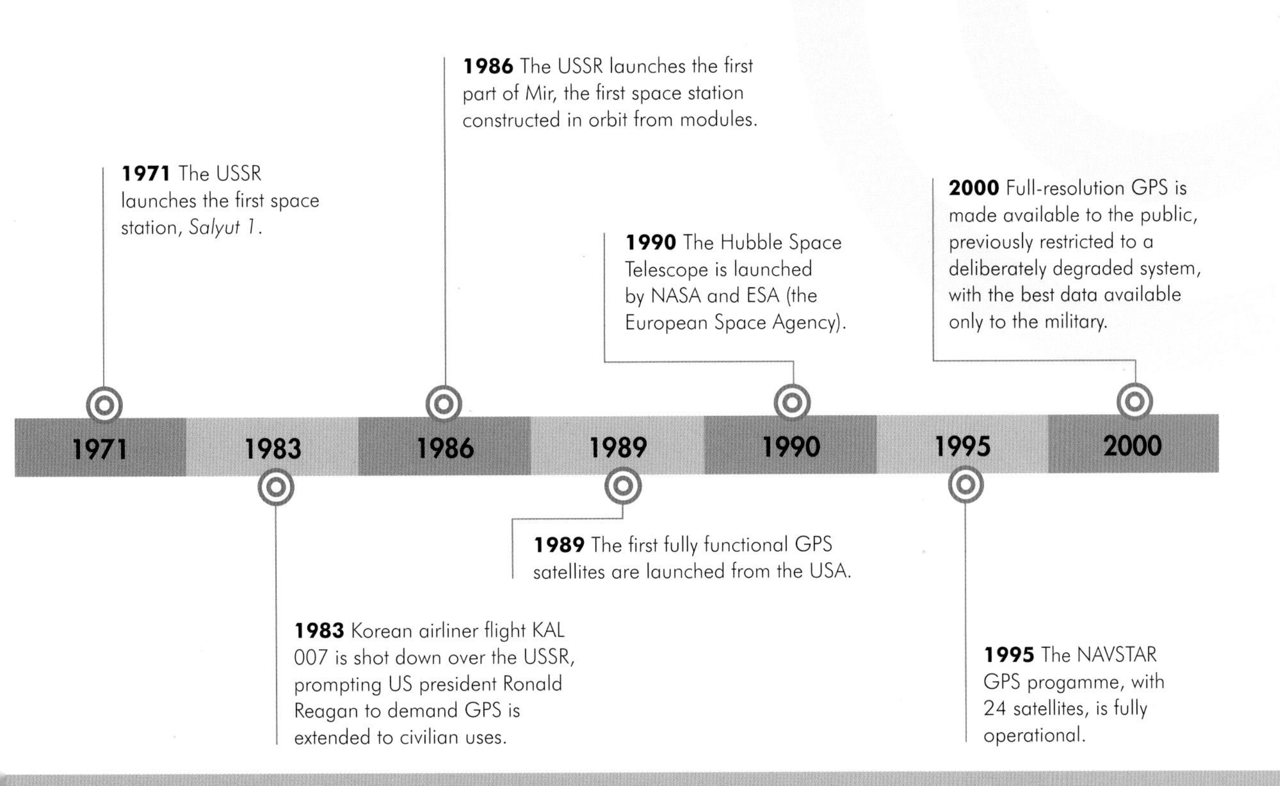

1990
SATELLITES AND SPACE

Some satellites are now helping us learn about deep space. One of the most famous is the Hubble Space Telescope, launched in 1990, a high-resolution optical telescope in low-Earth orbit (600km/372 miles above the Earth). It has taken the clearest photographs we have of distant galaxies, as being outside the atmosphere spares it from interference and background light. But it's not just for pretty pictures: Hubble has provided the data needed to calculate the rate of expansion of the universe. It is the only space telescope that is managed and repaired by astronauts who work from another type of satellite – space stations.

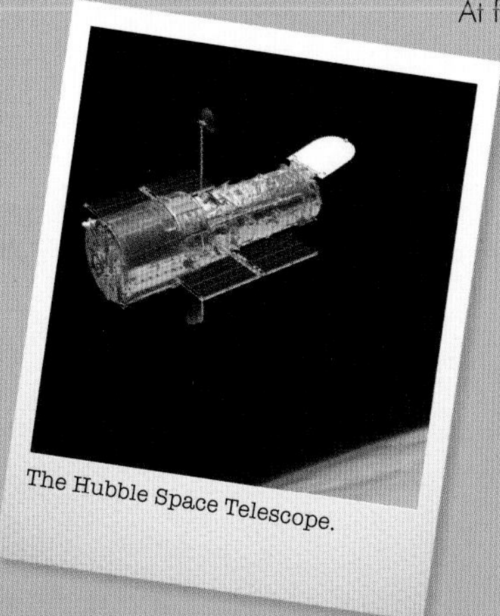
The Hubble Space Telescope.

1995
SATELLITES AND EARTH

Another of the now-ubiquitous uses of satellites is GPS (Global Positioning System). The algorithms used for GPS are simple: a position on Earth is triangulated with reference to three satellites of known location. The technique was developed the other way round, by two physicists trying to plot the path *Sputnik* had taken from changes in its radio signal. Their algorithms were reversed, years later, to locate a point on Earth from known satellite positions.

At first, GPS was used only for military purposes, but after a Korean airliner was shot down in 1983 after accidentally straying into Soviet airspace, it was opened up for civilian use. A fully functional system with 24 satellites covering the whole surface of the globe became operational in 1995.

1960–1979

The 1960s and 1970s were dominated by the Cold War and the space race that was a consequence of it. Ask anyone to name the most significant scientific moment of the 20th century and there is a slight chance they will cite the moon landings. But far from being a flashpoint, they were the culmination of the efforts of American and Soviet cosmonauts and engineers throughout the 1960s. The Apollo missions marked the end of ambitious exploration of space in manned vehicles – at least for the 20th century. Instead, attention turned to other methods of reaching out to the stars. The most ambitious and open-ended projects came in the 1970s, sending unmanned craft off into the unknown, carrying messages for any aliens that might encounter them.

For ordinary people, the most significant development of the 1970s was the advent of the personal computer. It's hard now to imagine life without computer technologies at our fingertips, yet when this period ended, the personal computer was only just beginning to creep into public consciousness, leaving the enclaves of garages in California and extending its reach beyond the geeks and hobbyists.

At the same time as optimism was rising in biology, with the unravelling of DNA offering the hope of new technologies and treatments, the first warning notes were sounded about the damage we are doing to the environment. They went largely unheeded, and it is only with hindsight that their importance looms large.

1961

CONQUERING SPACE: HUMAN SPACE FLIGHT

'I saw for the first time the Earth's shape. I could easily see the shores of continents, islands, great rivers, folds of the terrain, large bodies of water. The horizon is dark blue, smoothly turning to black...the feelings that filled me I can express with one word – joy.'

– Yuri Gagarin, 1961

Yuri Gagarin waves before his flight in *Vostok 1*.

In the midst of the Cold War, Soviet Russia scored a massive victory over the USA by putting the first human into space. The USA had already suffered a great blow when the USSR sent *Sputnik* into orbit, ahead of the USA's own satellite, in 1957. Just four years later, and after sending a sequence of flies, dogs and monkeys into space, the USSR launched Yuri Gagarin into space. His craft, *Vostok 1*, was a tiny spherical capsule only 2.3m (7.5ft) in diameter. It was blasted into space by a Vostok-K rocket. The entire flight, including a single orbit of the Earth, took only 108 minutes and remains the shortest-ever successful space flight.

The controls of *Vostok 1* were locked – everything was to be either automatic or controlled from the ground – and Gagarin did not need to do anything except be present. This precaution was taken because no one knew how weightlessness might affect the human body – whether Gagarin would be capable of flying the craft, making decisions or even surviving. Even so, he was given the code to unlock the controls just in case it became necessary. At the end of his flight, the craft's retroengine fired to drive it towards Earth and Gagarin ejected, parachuting to the ground in a field. He landed near a farmer and her daughter, who backed away in fear from this figure who had apparently (and actually) come from space. Following his uncomfortable landing, however, Gagarin became a national hero of the Soviet Union and Eastern Bloc, and a worldwide celebrity.

FLASHPOINT FACT

Even though the flight was scheduled to be over in just hours, *Vostok* carried a ten-day supply of food in case the retrorocket for the return to Earth failed. Gagarin, it was hoped, could survive on that until the orbit decayed and *Vostok* fell back to Earth.

TIMELINE

1951 The first dogs in space, launched by the USSR, are Dezik and Tsygan; both survive and return safely.

1957 The dog Laika is the first animal sent into orbit, on the USSR spacecraft *Sputnik 2*.

1957 The USSR launches the first satellite, *Sputnik*, into orbit around the Earth.

1959 The Soviet craft *Luna 2* crash-lands on the moon.

1965 Alexei Leonov performs the first tethered space walk from *Voskhod 2*.

1967 The first space fatality occurs, when Vladimir Komarov dies after the parachute of his *Soyuz 1* capsule fails to open on re-entry.

1951 1957 1959 1965 1967

FLASHPOINTS

1951
DOGS IN SPACE

Dogs became the first living creatures to be sent into space when Dezik and Tsygan flew to 110km (67 miles) in 1951. The first dog in orbit was Laika. She died a few hours into the flight from overheating. It was always known that she would die in space: there was no mechanism to bring *Sputnik 2* back to Earth. In all, there were 57 slots for dogs, and several for monkeys, during the 1950s and 1960s; some flew more than once. The dogs were all female (the dog spacesuit was designed to collect waste from female dogs) and were generally strays. They underwent a rigorous training programme to prepare them for the cramped and uncomfortable conditions.

Laika the dog went into space before Gagarin, in 1957.

1963
SPACEMEN AND WOMEN

After Gagarin's historic flight in 1961, the USSR sent several more cosmonauts into orbit, including the first woman in space, Valentina Tereshkova, in 1963. The USA was falling behind increasingly as the USSR also sent missions to Mercury and Venus. The remaining, ultimate target was to land a human on the moon. The *Apollo* programme was running behind schedule and it looked as through President Kennedy's promise to put a man on the moon before the end of the decade was going to be broken. The programme was accelerated, bringing forward the launch of *Apollo 8* to put astronauts in orbit around the moon at the end of 1968. The trip provided the first-ever view of the far side of the moon.

1968 *Apollo 8* orbits the moon, giving people on Earth the first ever view of the far side, which is always turned away from Earth.

1969 Two of the crew of *Apollo 11* land on the moon.

1971 The Soviet crew of *Soyuz 11* become the only deaths in space so far, dying after the decompression of their spacecraft during its return to Earth.

1986 The US space shuttle *Challenger* breaks up on take-off, killing the crew of seven.

2001 Dennis Tito becomes the first space tourist as a passenger, paying $20 million to spend eight days in space, six on the International Space Station.

1968 1969 1971 1986 2001

1969
MOON LANDING

The space race was won, ultimately, by the USA when Neil Armstrong stepped onto the surface of the moon on 20 July 1969. It is arguably the most important event in history – the first time a human has set foot on ground outside Earth. It marked the pinnacle of the technical achievements of the 20th century. Armstrong and Buzz Aldrin spent 21-and-a-half hours on the moon's surface taking photographs, collecting samples and conducting readings and experiments. Over the coming two-and-a-half years, 12 astronauts landed on the moon. Although there have been unmanned moon landings since, humans have not been back to the moon since 1972.

Post 1970
INFINITY AND BEYOND

While there have been no further moon landings, many astronauts have spent time in space in one of the space stations that were launched or built in space from the 1970s onwards – notably Skylab (manned 1973–4), Mir (1986–2001), the International Space Station (1998–present) and Tiangong (2011–present). These provide a base for astronauts staying in space for weeks or months, repairing important satellites and carrying out experiments and observations. Plans to go further afield – particularly Mars – are some way off. The stress that such a long flight (18 months or so) would present, and the difficulty of landing on and taking off from Mars, are among the challenges such a mission faces.

Buzz Aldrin walking on the surface of the moon.

1962

RACHEL CARSON: ENVIRONMENTAL CRUSADER

'Only within the moment of time represented by the present century has one species – man – acquired significant power to alter the nature of his world.'

– Rachel Carson

Rachel Carson.

Concern for the environment now drives an international agenda, yet before the work of Rachel Carson hardly anyone gave it a second thought.

Carson was born in 1907 and brought up on a farm in Pennsylvania where she developed a love for nature. After gaining a Master's degree in zoology, she became a science writer, working first for the US Fish and Wildlife Service writing radio broadcasts about marine life. She wrote many articles about wildlife and human interactions with nature, and several books. It was her last book, *Silent Spring*, that changed the world. Published in 1962, it publicized the harmful effects of the pesticides used extensively in farming at the time, and showed how they could permeate the food chain. Birds of prey were affected by feeding on smaller birds and fish in whose bodies pesticides had accumulated. The further up the food chain, the more DDT concentrated in an animal's body. Birds of prey suffered significantly, as they fed on animals that had often themselves fed on others that consumed plants sprayed with DDT. The birds laid eggs with shells too thin to survive, causing a drastic drop in numbers. The disaster brought the finely balanced web of interactions between organisms to light for the first time, revealing the ripple effects of disrupting the biome.

Carson was demonized by the agrochemical and farming industries, and endured personal and professional attacks. Even so, a Congressional hearing in 1963 led to a ban on the indiscriminate use of DDT, one of the most harmful of the chemicals she investigated. Although Carson died of cancer the following year, she had redefined humanity's relationship with the natural environment.

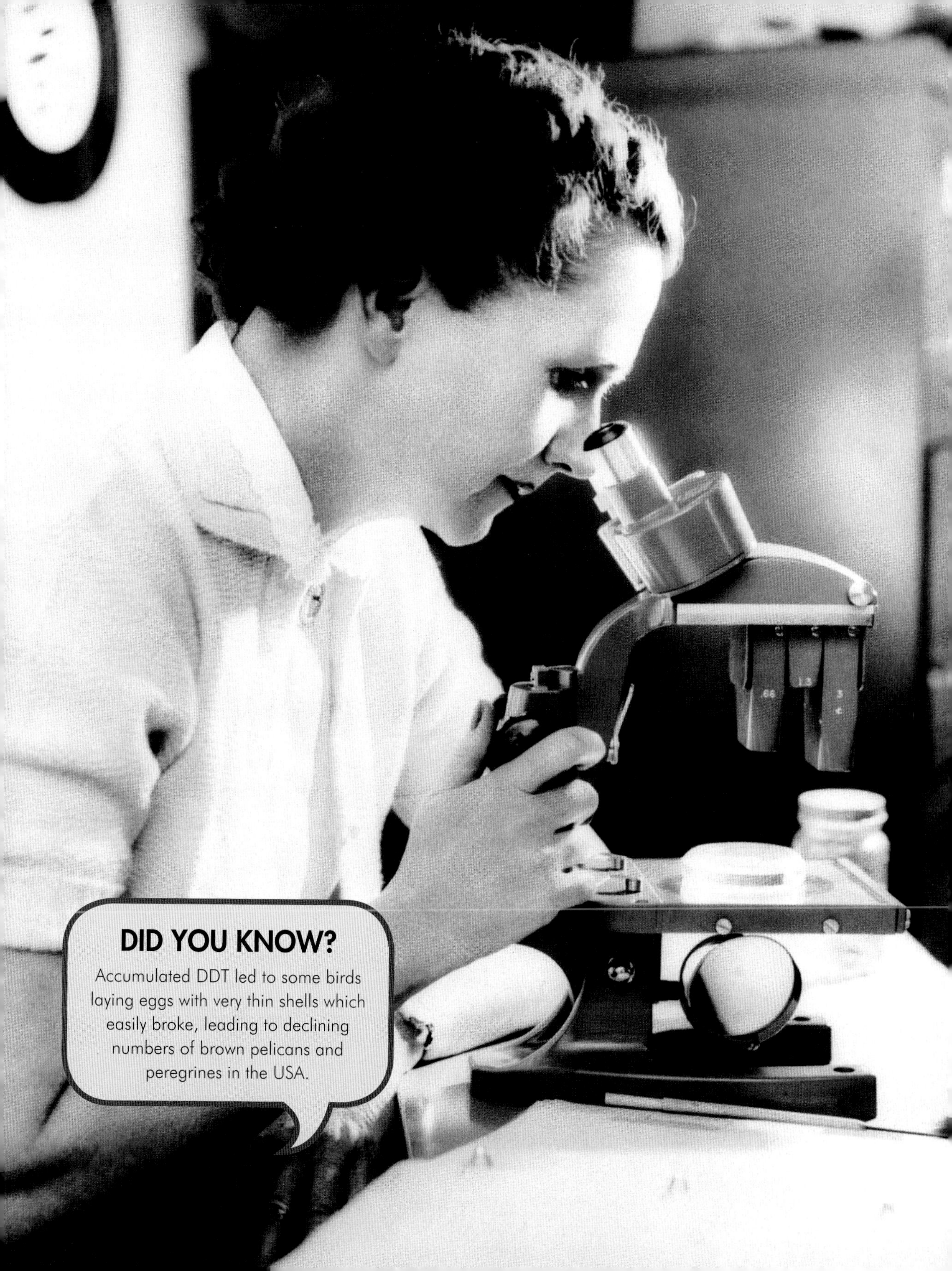

DID YOU KNOW?

Accumulated DDT led to some birds laying eggs with very thin shells which easily broke, leading to declining numbers of brown pelicans and peregrines in the USA.

TIMELINE

1921 George Washington Carver promotes soil conservation measures in North America.

1933 The start of 'dust bowl' conditions in North America and Canada caused by poor farming practices and extensive drought. Fierce winds whip the topsoil away, creating clouds of dust and leaving behind land unusable for farming.

1935 Carson joins the US Fish and Wildlife Service.

1939 The insecticidal properties of DDT are recognized and it is rapidly adopted in the battle against insect-borne disease.

1945 DDT is made available to farmers.

1947 The Los Angeles Air Pollution Control District is founded, the first board to try to control pollution in the USA.

1952 Carson wins the National Book Award with *The Sea Around Us*.

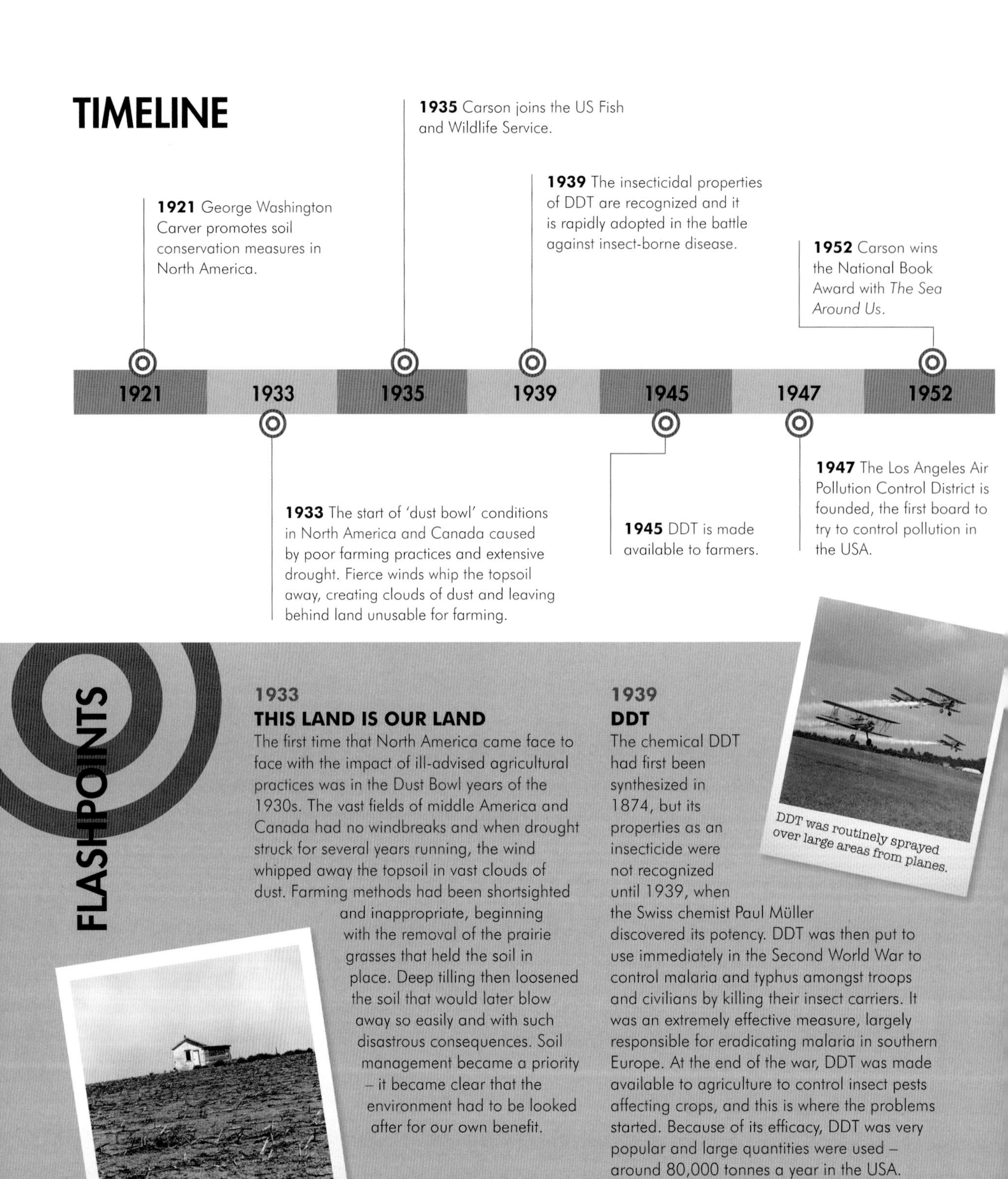

FLASHPOINTS

1933
THIS LAND IS OUR LAND

The first time that North America came face to face with the impact of ill-advised agricultural practices was in the Dust Bowl years of the 1930s. The vast fields of middle America and Canada had no windbreaks and when drought struck for several years running, the wind whipped away the topsoil in vast clouds of dust. Farming methods had been shortsighted and inappropriate, beginning with the removal of the prairie grasses that held the soil in place. Deep tilling then loosened the soil that would later blow away so easily and with such disastrous consequences. Soil management became a priority – it became clear that the environment had to be looked after for our own benefit.

The desolation of the Dust Bowl years led to farms being abandoned.

1939
DDT

The chemical DDT had first been synthesized in 1874, but its properties as an insecticide were not recognized until 1939, when the Swiss chemist Paul Müller discovered its potency. DDT was then put to use immediately in the Second World War to control malaria and typhus amongst troops and civilians by killing their insect carriers. It was an extremely effective measure, largely responsible for eradicating malaria in southern Europe. At the end of the war, DDT was made available to agriculture to control insect pests affecting crops, and this is where the problems started. Because of its efficacy, DDT was very popular and large quantities were used – around 80,000 tonnes a year in the USA.

DDT was routinely sprayed over large areas from planes.

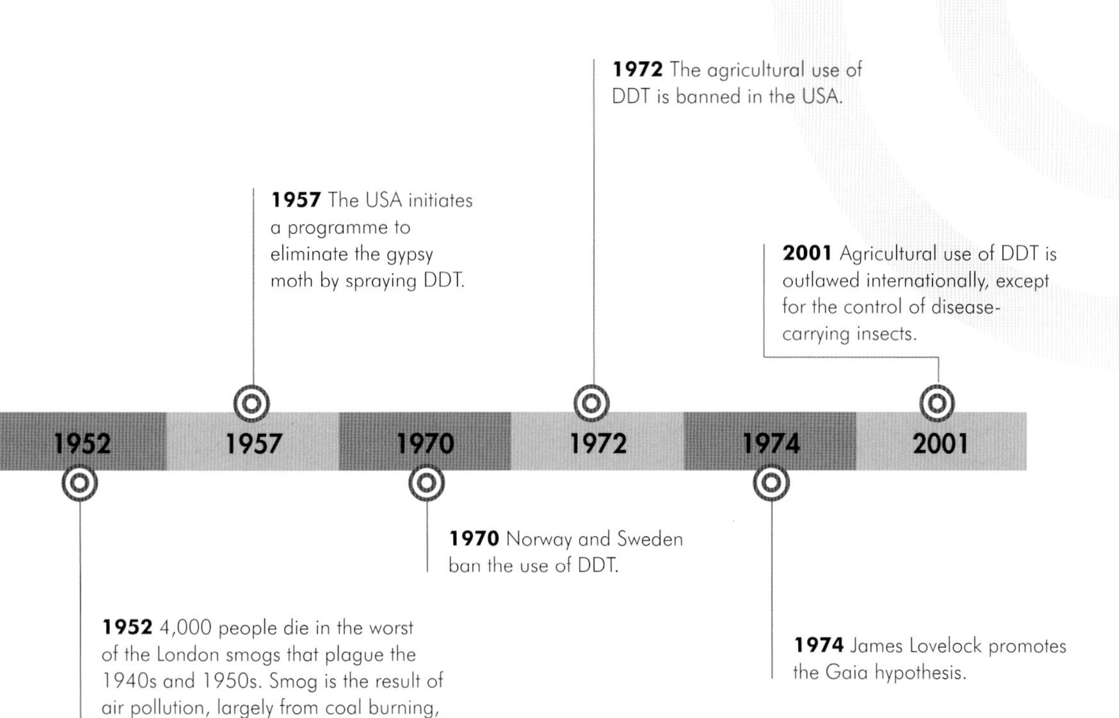

1957
THE PROBLEMS START

Some scientists voiced concern over the safety of DDT as early as the 1940s, but the public heard little of this until 1957, when an unsuccessful attempt was made to restrict the use of DDT in Nassau Country, New York. Carson heard about it, and was encouraged by *The New Yorker* to write an article, which eventually grew into *Silent Spring*. Carson discovered that DDT is a carcinogen (it can cause cancer), and that it was affecting wildlife beyond the targeted insects. In particular, DDT was killing fish and birds of prey in turn. The toxin slowly built up in eagles and peregrines that ate smaller birds which fed on insects or plants sprayed with DDT.

1974
THE EARTH AS LIVING BEING

The eclectic and eccentric scientist James Lovelock (1919–) proposed the Gaia theory in the 1970s: that the whole planet is self-regulating, and could be considered to be like a living organism. With Carson, Lovelock is considered one of the people who sparked the environmental movement and brought environmental issues to public attention. Gaia theory suggests that the Earth operates as a coherent and unified whole, which is generally self-healing – though the damage we are currently doing through global warming might have catastrophic effects. Although Lovelock is considered something of a maverick, and his theories are rejected by many scientists and ridiculed by some, his impact on public awareness, like Carson's, has been immense.

James Lovelock, 1997.

1962

AGAIN AND AGAIN: CLONES

'We shall escape the absurdity of growing a whole chicken in order to eat the breast or wing, by growing these parts separately under a suitable medium.'

– Winston Churchill, 1931

The African clawed frog, *Xenopus laevis*.

Dolly the sheep, the world's first cloned mammal.

The most famous clone is undoubtedly Dolly the sheep, born in 1996, yet she is not the most important in scientific terms. That honour goes to some humble tadpoles, produced in 1962 by John Gurdon at Oxford University.

Gurdon pioneered the process known as nuclear transplantation, which is used to create clones and stem cells. He built on work published by Thomas King and Robert Briggs in 1952 that explained how the nucleus of a cell from an early-stage embryo could be put into an unfertilized egg cell from which the nucleus had been removed and grown successfully. Gurdon's crucial innovation was to use an adult cell rather than an embryonic cell as the source.

Gurdon took egg cells from the African clawed frog *Xenopus laevis* and inserted into them the nuclei from tadpole gut cells. Even though the gut cells were already differentiated (that is, they were not stem cells but had grown into cells of a specific type), the eggs were able to develop into normal tadpoles. His success proved that the nucleus retains all the information needed to build a whole organism, even though much of that information is not actively used in the differentiated somatic (body) cell. This suggested potential for medical applications; if somatic cells could be taken from an individual and returned to the state of stem cells, they could perhaps be used to grow new tissue of different types.

Cloning produces an exact genetic copy of the original, single organism. This means that any cloned cells will be entirely compatible with the original donor – an important consideration in attempts to grow new tissue for transplants or grafts.

FLASHPOINT FACT
Dolly the sheep has been stuffed and is on display in the Royal Museum in Edinburgh, Scotland.

TIMELINE

1885 | 1901 | 1955 | 1962 | 1975 | 1986

1885 August Weismann proposes that as a cell differentiates, it loses genetic information.

1901 Hans Spemann splits a two-celled newt embryo in half and succeeds in growing two complete newts from the halves, refuting Weismann's theory.

1955 Thomas Briggs and Robert King report that by swapping the nucleus from an embryonic tadpole cell for the nucleus of an egg cell, they can grow a tadpole.

1962 John Gurdon successfully clones tadpoles from the African clawed toad, *Xenopus laevis*.

1975 César Milstein and Georges Köhler develop the technique for producing cloned antibodies.

1986 The first monoclonal antibody medicine is licenced for use.

1901
PART-WAY THERE

The first attempts at artificial cloning involved splitting early embryos in two. The German embryologist Hans Spemann experimented extensively with newt embryos. In one type of experiment, he grew two complete newts by splitting the first cells formed from a fertilized egg. This is what happens when identical twins form. In 1955 Thomas Briggs and Robert King took a crucial further step. They replaced the nucleus of a frog egg cell with the nucleus from a cell from a frog blastula, the ball of early cells produced as the fertilized egg divides. The egg grew into a normal embryo. But they had used undifferentiated cells; their experiment did not show whether genetic information is lost or changed once cells differentiate.

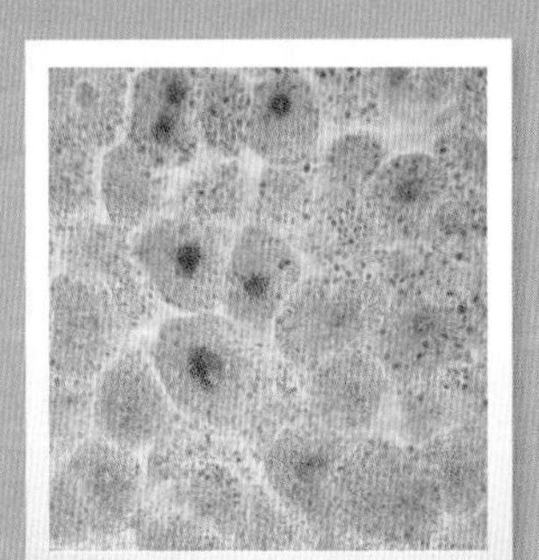

Part of a frog blastula, magnified x100.

1986
CLONED MEDICINE

The body fights infection by producing antibodies, each of which exactly fits the chemical profile of an attacking bacterium, virus or other foreign body. This means, potentially, that if we could produce the right antibodies outside the body we could use them to treat diseases. This is the principle of monoclonal antibodies. The first monoclonal antibody, Orthoclone OKT3, was registered for use in 1986; it is used as an anti-rejection drug for transplant patients. It was made by prompting mouse cells to grow clones of a particular type of antibody cell, so making a supply of identical antibodies that could be used as a medicine.

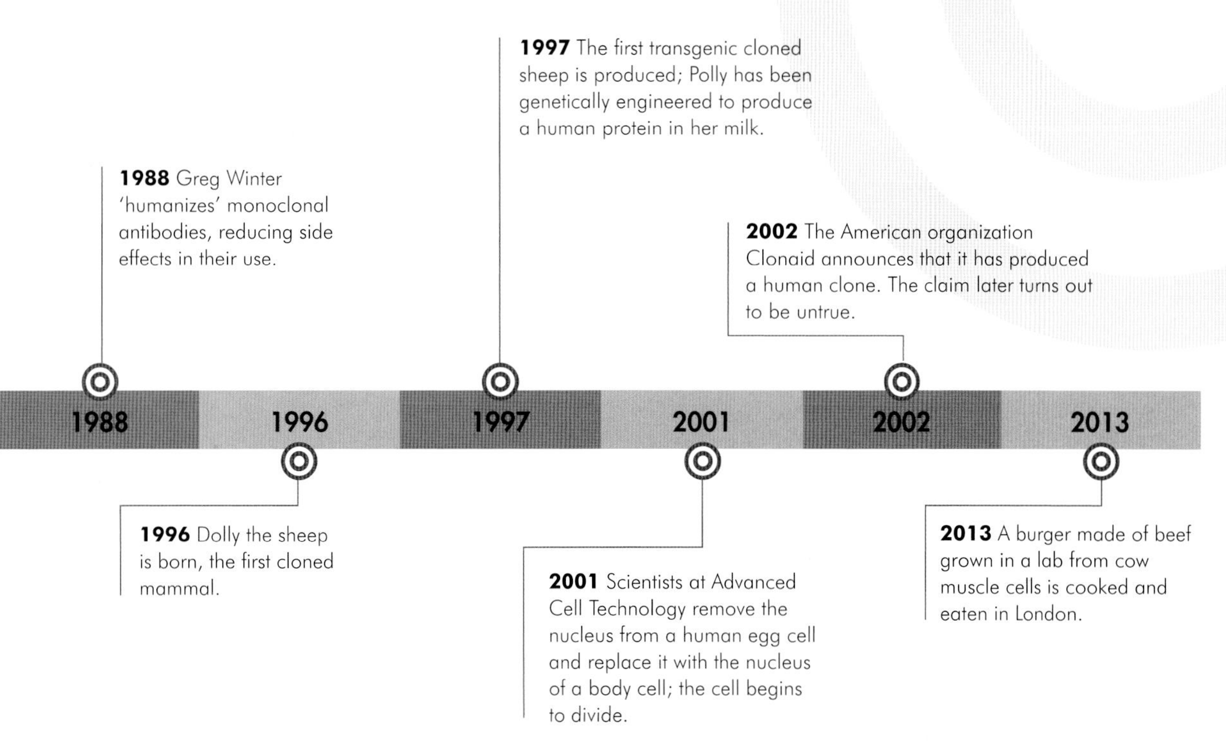

1996
DOLLY THE SHEEP

In 1995, researchers in Scotland produced two identical sheep from a single fertilized egg. The following year, Ian Wilmut created a cloned sheep from a somatic cell. His team took a cell from the udder of one sheep and put its nucleus into an egg cell from which the nucleus had been removed. The egg was then grown in a surrogate sheep-mother and born on 5 July 1996. The crucial detail of Dolly's heritage was that she had been produced from an adult somatic cell, not an embryonic or stem cell. This was final proof that no genetic information is lost in cell differentiation. We can clone an organism from a fully differentiated adult cell.

1997
PUTTING CLONES TO USE

The year after Dolly was born, another sheep made the headlines. Polly was not only a clone, but genetically engineered to produce a human protein in her milk, Factor IX, which prompts blood to clot. Polly sparked the new industry of 'pharming' – producing medicines and possibly even transplant organs by prompting animals to express human genes. Cloning is also used to reproduce high-yielding food animals and plants. By cloning muscle tissue, it might eventually be possible to grow meat without growing animals, as Churchill predicted. The first burger made from beef grown in a lab was shown in 2013.

The world's first laboratory-grown beef burger.

1967
CLIMATE CHANGE

'At the moment we cannot predict what the overall climatic results will be of our using the atmosphere as a garbage dump.'

– Paul R Ehrlich, *The Population Bomb*, 1968

The polar bear has become a familiar icon of climate change.

The idea that gases such as carbon dioxide (CO_2), water vapour and methane might produce a greenhouse effect on Earth was first suggested in 1896. It only came to be considered a serious threat from 1967, though. In that year, the Japanese climatologist Syukuro Manabe used computers and a more complex model to re-examine the link between rising temperatures and the concentration of carbon dioxide in the atmosphere. His first trials found that doubling the CO_2 in the atmosphere would cause a rise of only 2°C. Manabe revised his model to take account of the potential of the oceans to act as a heat sink, deriving the first global ocean-coupled, climate-modelling program. The calculations were so complex that, using a computer with only 500Kb of memory, it took 50 days to run the program and return the results.

Full climate modelling is extremely complex, as it has to take account not just of the heating effect of greenhouse gases but of all the ensuing consequences that then feed back into the model. The oceans can absorb heat, as a vast, dark body. As the temperature of the sea rises, ice in and bordering the sea melts, and as the temperature of the air rises, ice on land melts. The melting ice adds water to the oceans, raising the sea level. Rising sea levels mean that more land is underwater, and so the surface area of the ocean is larger. This simultaneously increases the area of dark surface and removes areas of ice that previously helped to reflect heat back from the surface of the Earth. Consequently, melting ice and rising sea levels contribute to an ever-increasing speed of heating. At the same time, removal of forest and other greenery adds to the build-up of CO_2, as its removal through photosynthesis is reduced. Rising temperatures lead to desertification, making land inhospitable to plant life and so further reducing photosynthesis. This, too, fuels an escalating spiral of rising temperatures and raised CO_2 levels. Climate scientists are now almost unanimous in considering human activity to be driving high levels of CO_2 and consequent climate change.

FLASHPOINT FACT

The paper reporting Manabe's first results on global warming is one of the most frequently cited scientific papers in history.

TIMELINE

1896	1938	1956	1958	1960	1967	1968	1973

1896 Svante Arrhenuis proposes a link between rising global temperatures and increasing CO_2 in the atmosphere.

1938 Guy Callendar argues that CO_2-triggered global warming is already happening.

1956 Ewing and Donn suggest a rapid-onset ice age is possible.

1958 It is discovered that a greenhouse effect on Venus has raised the surface temperature to well above the boiling point of water.

1960 Charles Keeling accurately measures the CO_2 in the atmosphere and finds it is rising year on year.

1967 Syukuro Manabe calculates the expected rise from doubling of CO_2 levels as 2°C.

1968 Studies suggest that Antarctic ice sheets could collapse, triggering a catastrophic rise in sea levels.

1973 James Lovelock suggests that CFCs (chlorofluorocarbons) could have a greenhouse effect 10,000 times worse than CO_2.

1896
HOT OR COLD?

A link between atmospheric CO_2 and climate change was first proposed by Svante Arrhenuis in 1896. He set out to study the cause of the ice ages, and focused on the effects of CO_2 and water vapour trapping heat near the Earth's surface and acting like a greenhouse. Working before the age of computers, he carried out tens of thousands of calculations to reach his conclusion that halving the level of CO_2 would cause a temperature drop of 4–5°C, and doubling it would cause a rise of 5–6°C. From coal consumption at the time, he estimated that it would take 3,000 years to increase CO_2 by 50 per cent and would bring pleasant, mild conditions.

1960
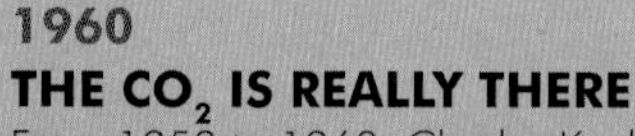

THE CO_2 IS REALLY THERE

From 1958 to 1960, Charles Keeling made crucial measurements at the Mauna Lao Observatory on Hawaii. He found that atmospheric CO_2 follows a seasonal pattern, rising in the northern hemisphere's winter and falling in the summer, but also showing an overall rise over the years. The seasonal pattern is accounted for by use of fossil fuels rising in the winter (for heating) and the uptake of CO_2 by plants being lower. In spring, uptake by plants begins to rise, and the level of CO_2 begins to fall. More important was the trend of an overall rise. Levels of CO_2 have now been mapped continuously since the start of Keeling's study and show a steady and accelerating rise, year on year.

Human dependence on fossil fuels makes climate change difficult to combat.

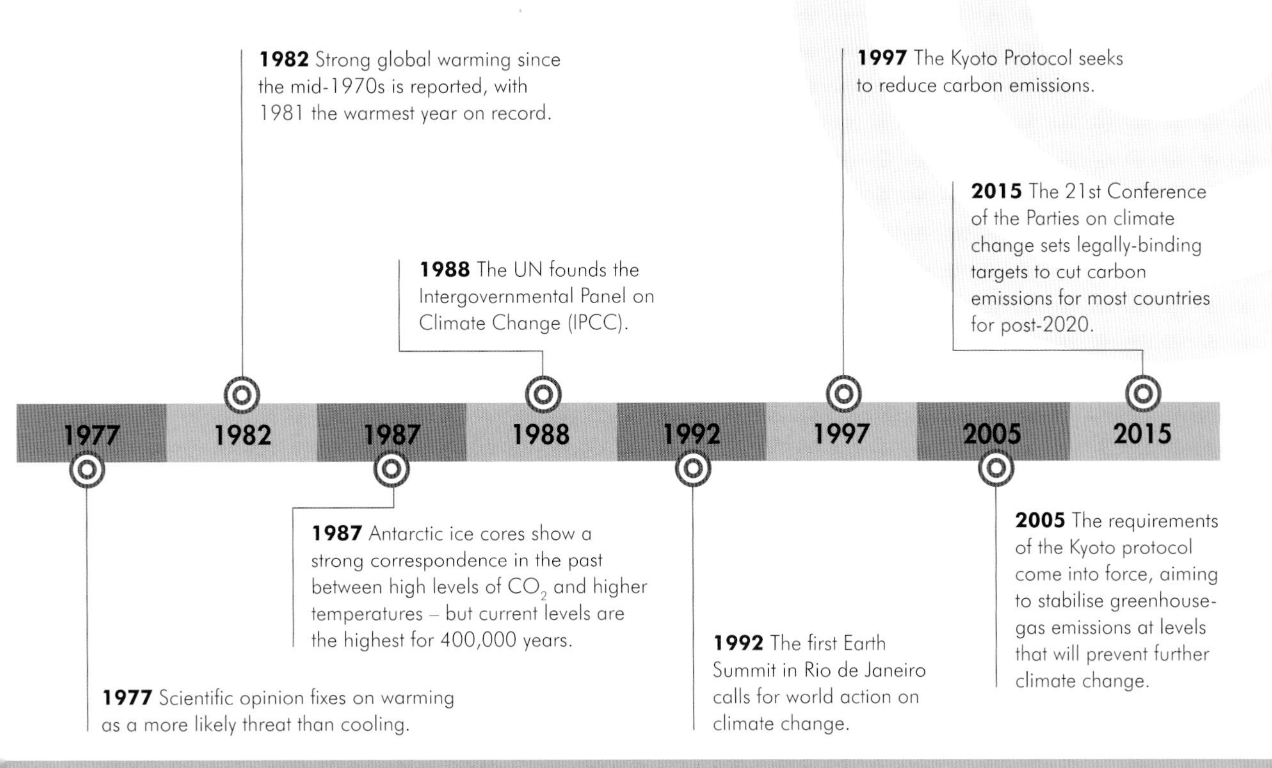

1978
FROM DISSENT TO CONSENSUS

There were plenty of dissenters in the early years. At first, people said the oceans would easily absorb any additional CO_2. Then there was the timescale – early models suggested that any rise in temperature would be very gradual, taking centuries. There were also plenty of people, both scientists and others, who thought that the warming seen over the previous century could have another cause, such as solar activity. Evidence from ice cores, which trap bubbles of air for hundreds of thousands of years, enabled scientists in the 1980s to analyze the atmosphere of the past. In 1978 it emerged that the level of CO_2 in the atmosphere was unprecedented, and steadily rising.

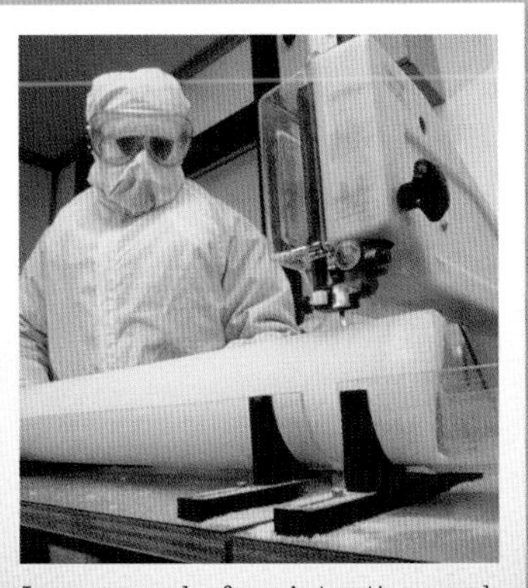
Ice core samples from Antarctica reveal the composition of the atmosphere hundreds of thousands of years ago.

1997
TIME FOR A CHANGE

By the end of the 1980s, most scientists were convinced that there is a real problem. The Earth Summit in Rio de Janeiro, Brazil, in 1992 brought together representatives of 172 governments to tackle the problems of damage to the environment and climate change. It laid the groundwork for the Kyoto Protocol, which, in 1997, set out requirements for countries to limit carbon emissions. The target was to reduce greenhouse gas emissions to 5.2 per cent below their 1990 levels by 2012 in the developed countries. No country has achieved that target; CO_2 levels continue to climb and are now over 400ppm (parts per million).

1973
BUILD-YOUR-OWN ORGANISMS: RECOMBINANT DNA

'Recombinant DNA was the most monumental power ever handed to us. The moment you heard you could do this, the imagination just went wild.'

– David Baltimore, ex-president of the California Institute of Technology

Gene splicing: plasmids (yellow) with different gene sequences (various colours) spliced into them.

Genetic engineering involves working with the material of inheritance – DNA – to manage the characteristics of an organism. In its least invasive form, it is selective breeding – choosing the plants or animals with desirable features and breeding from them to strengthen those features. With the work of Herbert Boyer and Stanley Cohen in 1973, though, manipulation moved within the cell.

Boyer and Cohen developed the technique of recombining DNA to fuse chunks of DNA from one organism into the DNA of another. To do this, they used an enzyme to cut the DNA of the host organism, then introduced the required bit of DNA from the source organism between the cut ends. They used an enzyme again to prompt the DNA to reconnect around the new segment. Their first trial involved cleaving the DNA of the bacterium *E. coli* and adding a gene that provides resistance to the antiobiotic kanamycin. Future generations of *E. coli* grown from the spliced bacteria continued to show resistance to kanamycin – the bacteria had made the gene their own, passing it from one generation to the next. Further experiments succeeded in introducing a toad gene into bacteria showing that, in principle, rapidly reproducing bacteria could be used to manufacture products associated with more sophisticated organisms.

This is potentially extremely useful, as it means that proteins with important medical or industrial uses can be produced quickly and easily. One of the first and most significant uses of recombinant DNA technology was to create bacterial 'factories' making human insulin to treat people with diabetes. Previously, insulin supplements were derived from slaughtered animals – a much slower process with additional medical complications.

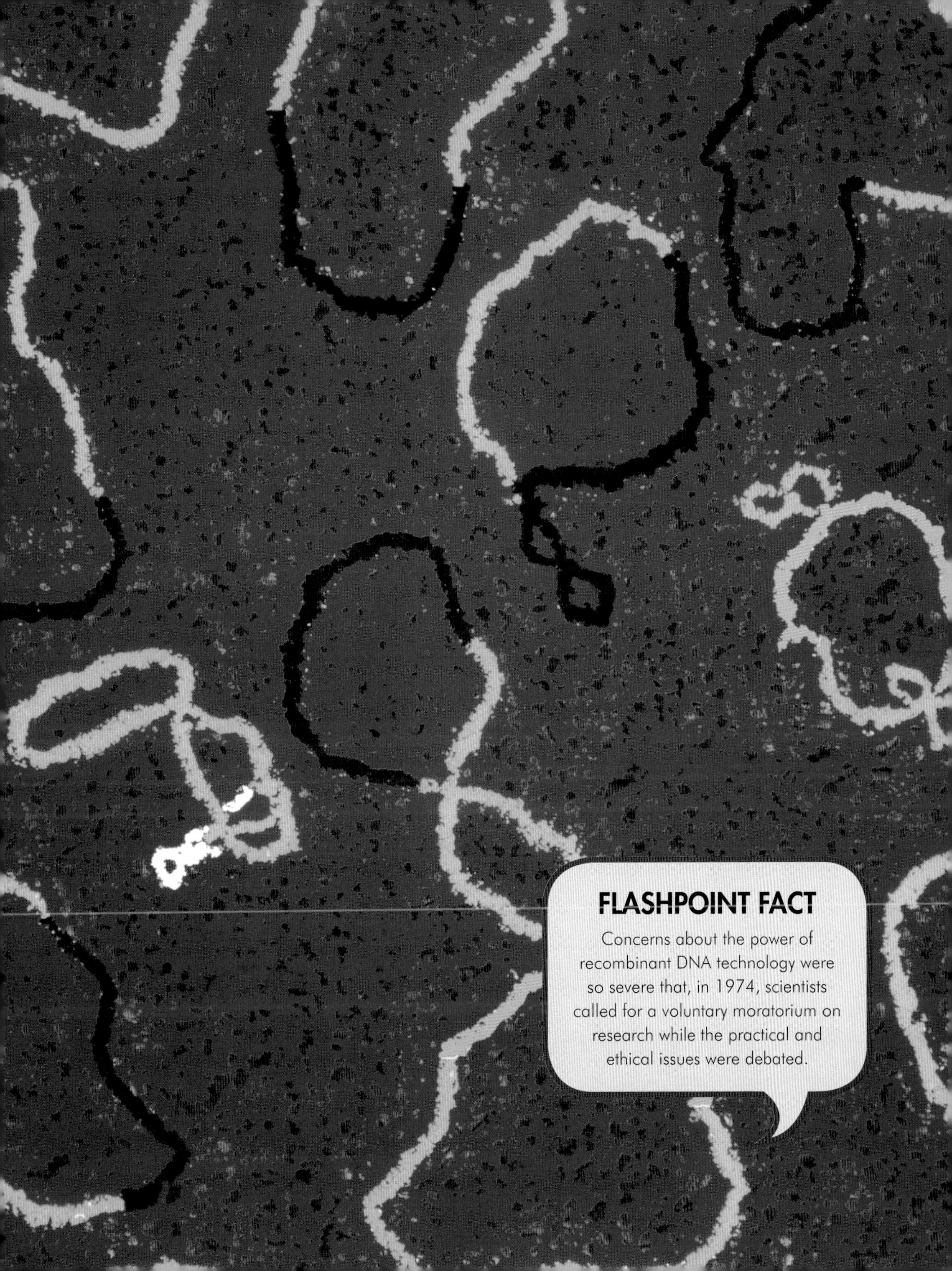
FLASHPOINT FACT
Concerns about the power of recombinant DNA technology were so severe that, in 1974, scientists called for a voluntary moratorium on research while the practical and ethical issues were debated.

TIMELINE

1944 Oswald Avery discovers that DNA can transfer genetic information between organisms.

1952 The Hershey-Chase experiment confirms that DNA is the means of inheritance.

1953 The structure of DNA is discovered by Francis Crick, James Watson and Rosalind Franklin.

1960 Sydney Brenner, Francis Crick, François Jacob and Jacques Monod discover the mechanism by which messenger-RNA (mRNA) facilitates the production of proteins in a cell from the instructions carried in DNA.

1972 Paul Berg discovers restriction enzymes, which can be used to cut DNA at a specific point.

FLASHPOINTS

1945–75

QUESTIONS OF SAFETY AND ETHICS

Berg was not the only one to worry. There has been considerable public opposition to some research in and products of GM technologies. Early on, in 1974, scientists in the USA called a voluntary halt to development while the safety of using recombinant DNA was discussed. They held a conference in 1975, where it was decided to proceed with caution. Consumer resistance to GM foods has meant a considerably reduced market for it in the EU, for instance. Although apparently safe for consumers to eat, GM foods raise concerns about ethics and animal welfare, the safety of the environment (the possibility of GM organisms contaminating natural populations), and unforeseen consequences.

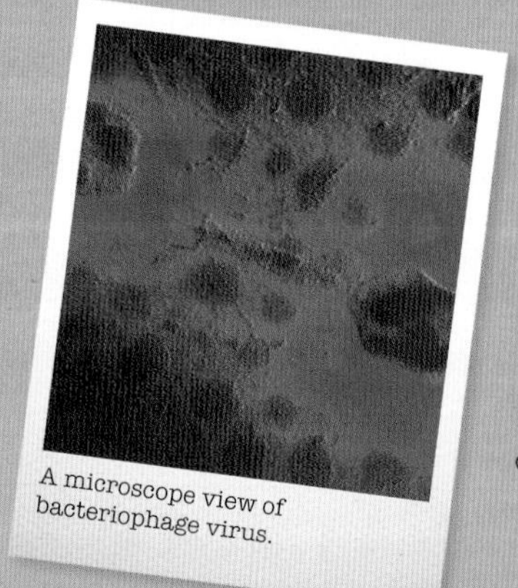

A microscope view of bacteriophage virus.

1972

CUT AND SPLICE

Paul Berg set the ball rolling in 1972 when he discovered that restriction enzymes will cut through DNA. Working with viruses that have a closed loop of DNA (called a plasmid), he opened the loops, then added another enzyme to make the cut ends 'sticky' and receptive. He added the new piece of DNA and closed the loop, now with DNA from two viruses combined. The resultant virus had features of both the originals. One virus was a bacteriophage – a virus that multiplies inside bacteria. Berg had planned to put it into *E. coli* and use the bacterium to grow the new virus. In the event, he decided it was dangerous to risk the escape of a common bacterium carrying a manufactured virus, and deferred that part.

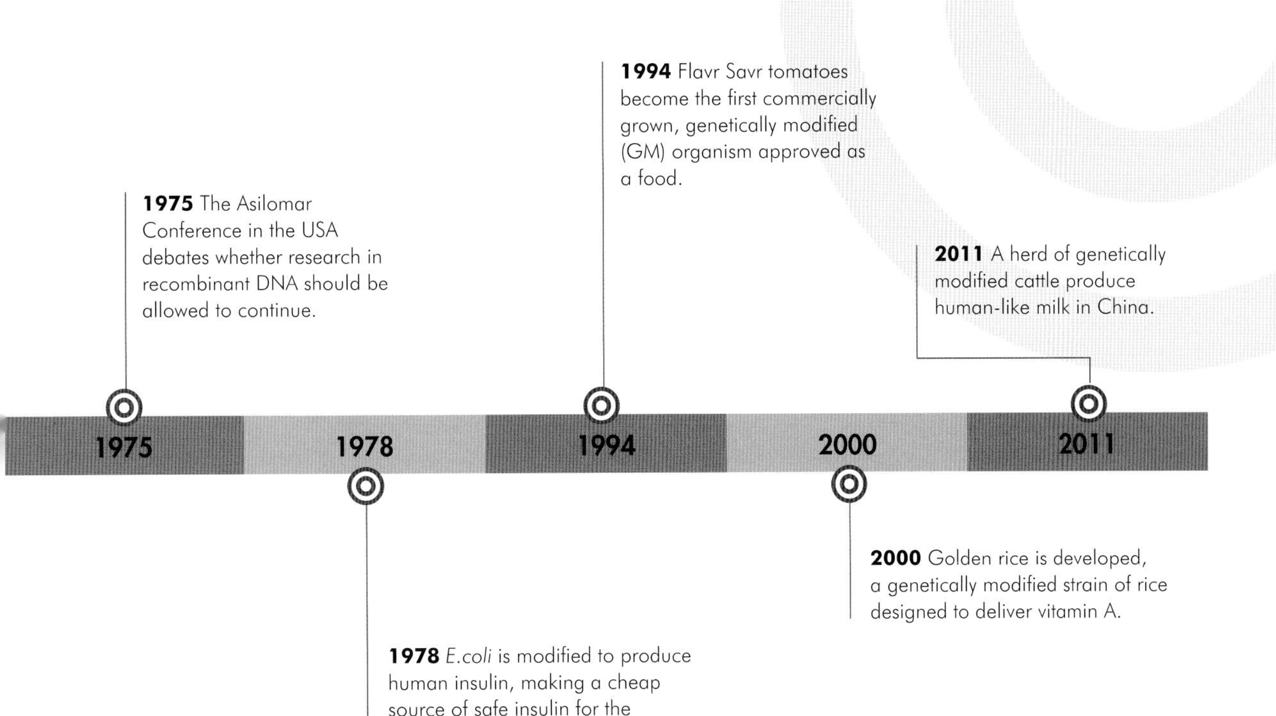

1978
BIOTECH STARTS HERE

The biotechnology industries work with naturally occurring organisms and life processes to make new products. One of the early successes, in 1978, was the production of human insulin (for treating diabetes) by splicing a gene into a carrier plasmid and inserting it into *E. coli* bacteria. Growing the bacteria then produced biological insulin 'factories'. Insulin is now produced in giant vats and is entirely human-compatible, whereas previously it had been collected from animals slaughtered for meat. Other developments include modifying mammals (such as sheep and cows) to produce specific chemicals in their milk. One ambitious goal is to produce a good mimic of human breastmilk from a cow, goat or sheep, with some success reported in China in 2011.

1994
GM FOODS

The use of genetic engineering to produce foods that have been designed to keep better, to be grown more economically or to confer additional nutrional benefits began in 1994 with the approval of the genetically modified tomato Flavr Savr. By the addition of a gene to interfere with the production of an enzyme that ripens tomatoes, the Flavr Savr was made to last longer without softening. Other changes in GM foods make them more resistant to disease, to frost or to insect pests. The rice strain 'golden rice', developed in 2000, includes a gene that leads it to produce a precursor to vitamin A and was heralded as a way of combatting vitamin-A deficiency, particularly in parts of Africa and in India.

A genetically engineered Flavr Savr tomato.

1974

PUTTING THE ATOM TOGETHER: THE STANDARD MODEL

'I have heard it said that the finder of a new elementary particle used to be rewarded by a Nobel Prize, but such a discovery now ought to be punished by a 10,000 dollar fine.'

– Willis Lamb, Nobel Prize acceptance speech, 1955

The elementary particle 'zoo'.

In 1974, John Iliopoulos tried to put in order all the elementary particles and fundamental forces that had been discovered over recent years. The complexity of the atom had greatly increased during the course of the 20th century – or, at least, physicists' knowledge of it had increased. At the start of the century, the only elementary particle known was the electron. The neutron and proton were discovered in the early decades, but turned out not to be elementary after all – they could be further broken down.

Iliopoulos's framework became known as the Standard Model. It proposes that elementary particles interact by means of the fundamental forces to make up all the matter of the universe. The forces are the weak and strong interactions, and electromagnetism. The strong interaction operates only at a very small scale but is responsible for the cohesion of matter, as it holds quarks together. The weak interaction is responsible for radioactive decay and nuclear fission. The particles are divided into bosons, which are carriers of the forces, and fermions, which are components of matter.

Even so, the Standard Model falls short of being a complete theory of fundamental interactions. It does not account for dark matter or dark energy, and it doesn't fully incorporate the theory of gravitation, as described by Albert Einstein's General Relativity. Also missing is an explanation for the continued expansion of the universe – something scientists are still working on today. Because of its success in explaining a wide variety of experimental results, the Standard Model is sometimes regarded as a 'theory of almost everything'.

DID YOU KNOW?

Gell-Mann came up with the name 'quark' before he had decided on a spelling, but knew he wanted it pronounced 'kwork'. He found the spelling in Joyce's *Finnegans Wake*: 'Three quarks for Muster Mark!'.

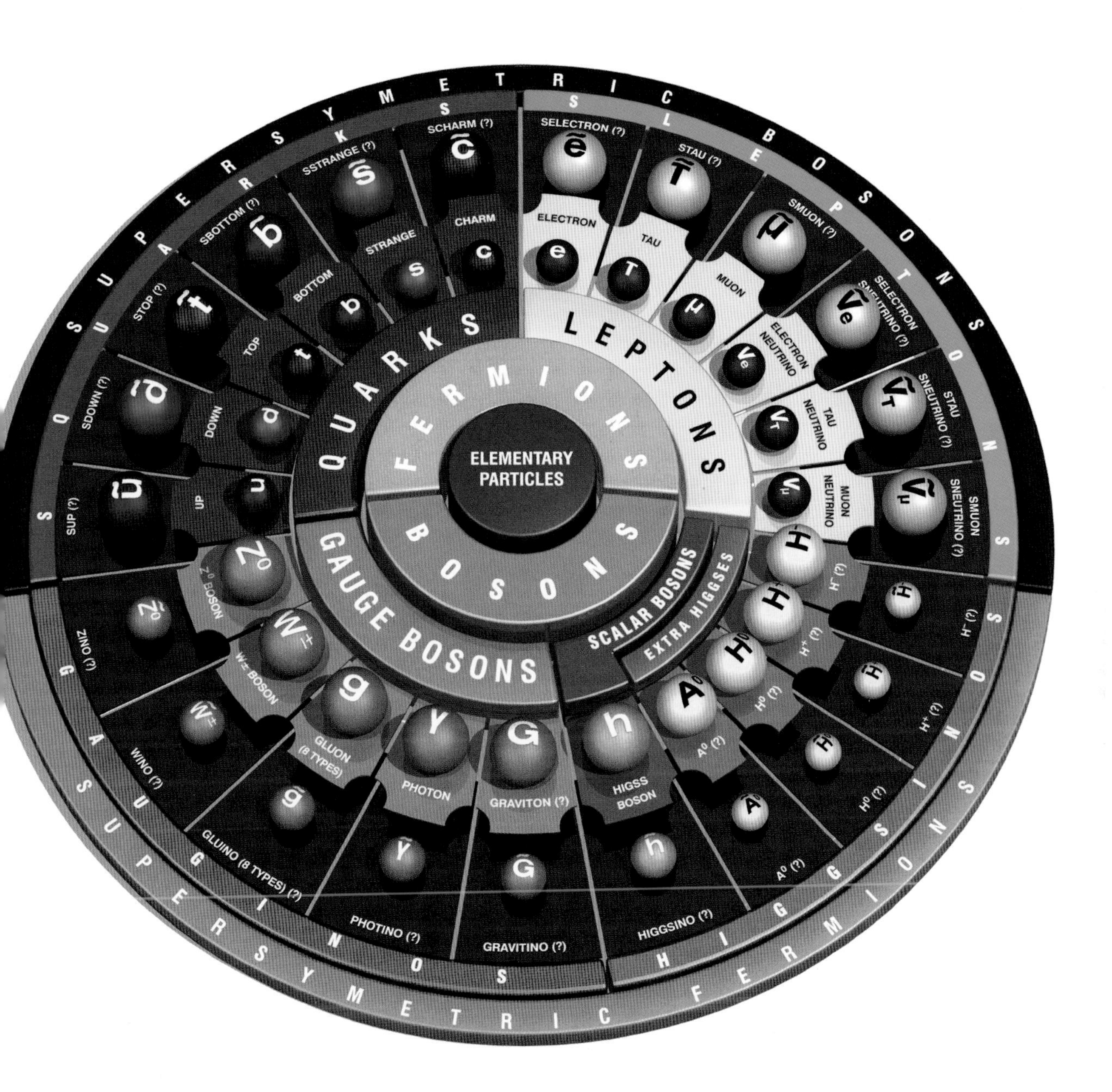
ELEMENTARY PARTICLES
FERMIONS
BOSONS
QUARKS
LEPTONS
GAUGE BOSONS
SCALAR BOSONS
EXTRA HIGGSES
UP
DOWN
TOP
BOTTOM
STRANGE
CHARM
ELECTRON
TAU
MUON
ELECTRON NEUTRINO
TAU NEUTRINO
MUON NEUTRINO
Z BOSON
W BOSON
GLUON (8 TYPES)
PHOTON
GRAVITON (?)
HIGGS BOSON
SUP (?)
SDOWN (?)
STOP (?)
SBOTTOM (?)
SSTRANGE (?)
SCHARM (?)
SELECTRON (?)
STAU (?)
SMUON (?)
SELECTRON SNEUTRINO (?)
STAU SNEUTRINO (?)
SMUON SNEUTRINO (?)
ZINO (?)
WINO (?)
GLUINO (8 TYPES) (?)
PHOTINO (?)
GRAVITINO (?)
HIGGSINO (?)
SUPERSYMETRIC BOSONS
SQUARKS
SLEPTONS
SUPERSYMETRIC FERMIONS
GAUGINOS
HIGGSINOS

TIMELINE

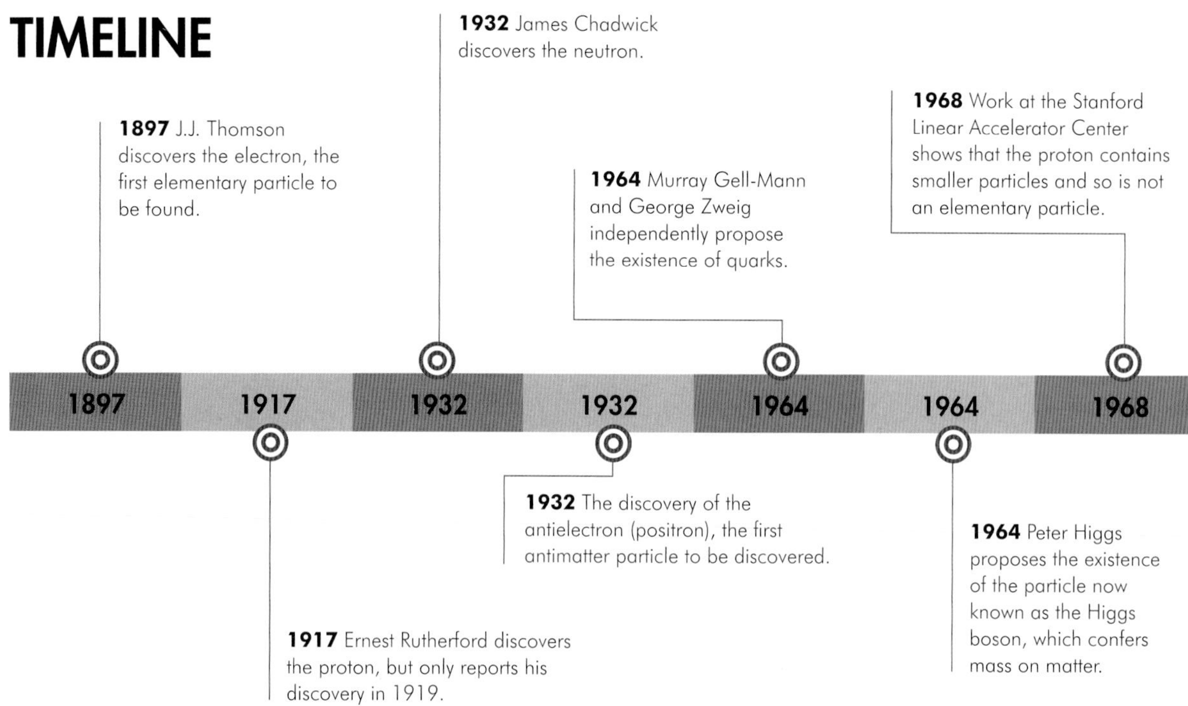

1964
HIGGS AND HIS BOSON

The Higgs mechanism was proposed in 1964 to explain why some bosons, which should theoretically be massless, appear to have mass. The idea is that an unusual type of field, called the Higgs field, exists all through space. The Higgs field comes with its own particle, the Higgs boson, which interacts with the field – or, more accurately, is a perturbation in it – to confer mass on other particles. It became clear later that the Higgs mechanism could also explain why electrons and quarks have mass. As all baryonic (ie. normal) matter is made of electrons and quarks, this makes the Higgs mechanism responsible for the mass of matter. The existence of the Higgs boson – which self-destructs after only 10^{-22} second – was confirmed in 2013.

1968
QUARKS

An experiment at the Stanford Linear Accelerator Center (SALC) in 1968 revealed that protons are not elementary particles after all, but are made up of smaller particles. Murray Gell-Man proposed the name 'quark' for these, initially suggesting that there were three. The model was later extended to six. They have the odd names 'up', 'down', 'charm', 'top', 'bottom' and 'strange'. Each proton and neutron is made up of three quarks, and the quarks are held together by gluons. Gluons are massless bosons that mediate the strong interaction. The other quarks were slower to find. The top quark, found last of all in 1995, has the mass of a tungsten atom and is highly unstable. It can be generated only in high-speed particle accelerators and lasts only 5×10^{-25} second.

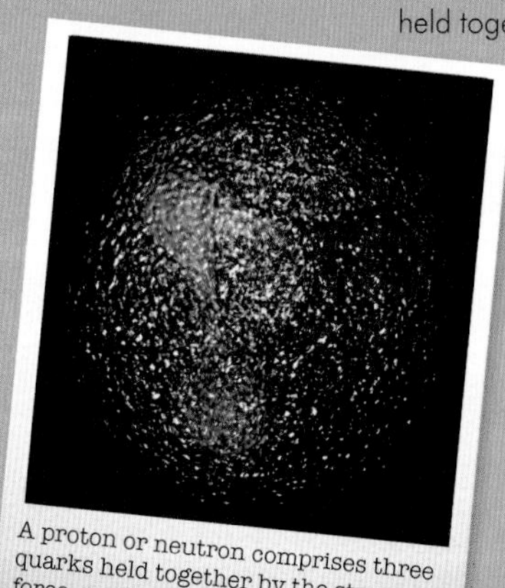

A proton or neutron comprises three quarks held together by the strong force, mediated by gluons.

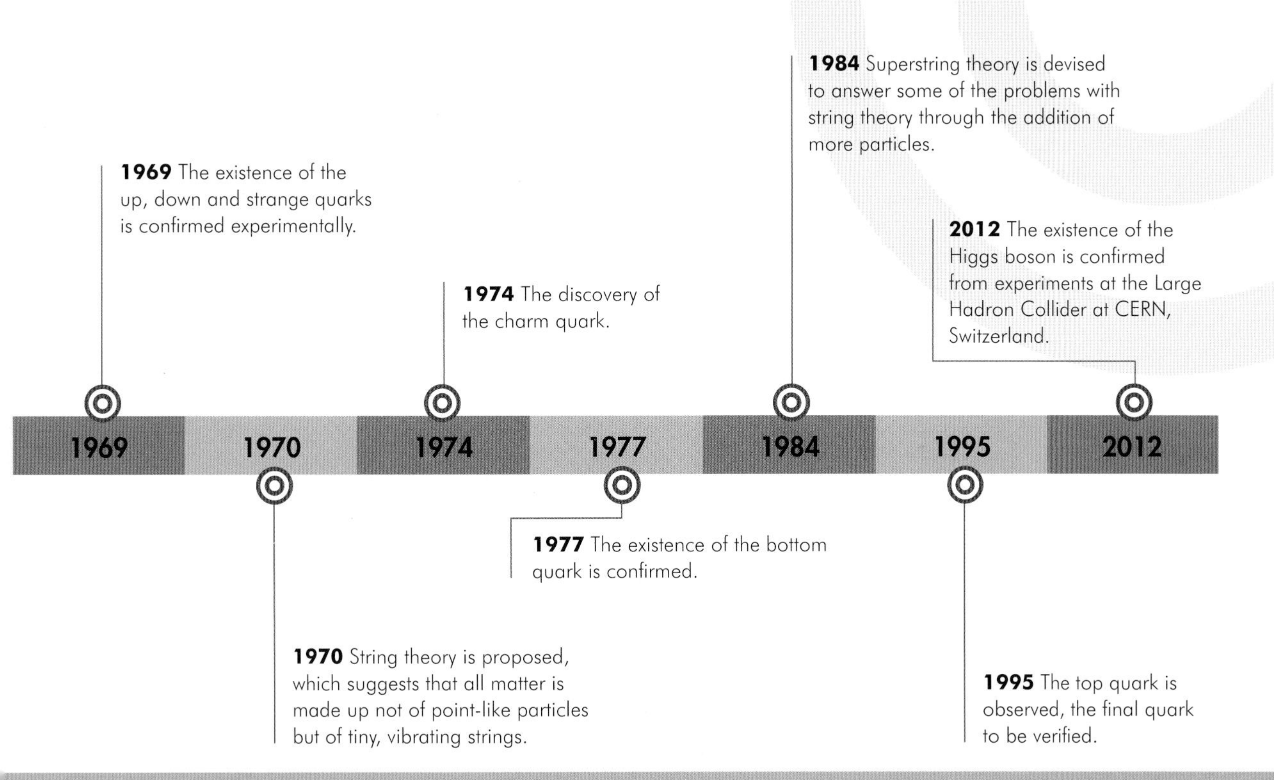

1970/1984
STRINGS AND SUPERSTRINGS

The Standard Model can account for three of the four forces, but not gravity. In 1970, Larry Suskind suggested string theory that could, if correct, wed gravity to the other forces and matter. Instead of seeing matter as comprising tiny, point-like particles that essentially occupy no dimensions, he suggested that they might be tiny, vibrating strings, with extension in one dimension but folded into ten dimensions. The difference between particles is explained as differences in vibration patterns. Experimental testing did not uphold string theory, and it fell from favour. It was resuscitated in 1984 in superstring theory, supplemented with extra particles so that each particle had a partner – an electron paired with a selectron, for example. Superstring theory cannot be tested experimentally.

Pre-1974
THE PARTICLE ZOO

The particle zoo was an informal name given to the huge number of particles known or theorized before the categorization of the Standard Model in 1974. It includes both elementary and composite particles (those made up from elementary particles). They all have their particular roles to play, and they all have associated anti-particles. The Standard Model brought some order to the zoo – a bit like Carl Linnaeus's attempt to categorize living organisms in the 1730s. It imposes a framework in which similarities and differences, relationships and roles, become easier to understand.

The inside of the Hadron Collider, the world's largest particle collider.

1974

THE STORY OF HUMANKIND: LUCY

'In our case, finding a Lucy is unique. No one will ever find another Lucy. You can't order one from a biological supply house. It's a unique discovery, a unique specimen.'

– Donald Johanson

The remaining fossilized bones of Lucy.

On 24 November 1974, Donald Johanson and Maurice Taieb made an astonishing discovery in a ditch in Ethiopia. They recognized immediately that the bones they had found belonged to an early human ancestor. These turned out to fragments of a skeleton of *Australopithecus afarensis*, and the most important proto-human remains ever found. The skeleton, later named Lucy, is 3.2 million years old, pushing early human evolution back by 400,000 years and siting it in Africa. The discovery of Lucy was particularly exciting because 40 per cent of her bones were recovered, including parts of the ribs and skull, which rarely survive. Lucy spent most of her time walking upright, a distinctive characteristic of humans, and is the earliest we know to have done so. She might also have spent time in the trees; she had long, trailing arms like a baboon. Her teeth are between those found in other apes and human teeth; her brain was smaller than a human brain. For a while, it looked as if Lucy might represent the most recent common ancestor of both humans and chimpanzees.

An example of *A. afarensis* had been found in 1924, the Taung Child, but was dismissed as an ape of no significance. By the time Lucy was discovered, the importance of *A. afarensis* was recognized. Although older and more complete fossils of *A. afarensis* have been found since, Lucy sparked a new era of investigation into human ancestry.

FLASHPOINT FACT

Lucy was named after the Beatles' song, 'Lucy in the Sky with Diamonds', played in the camp after her discovery.

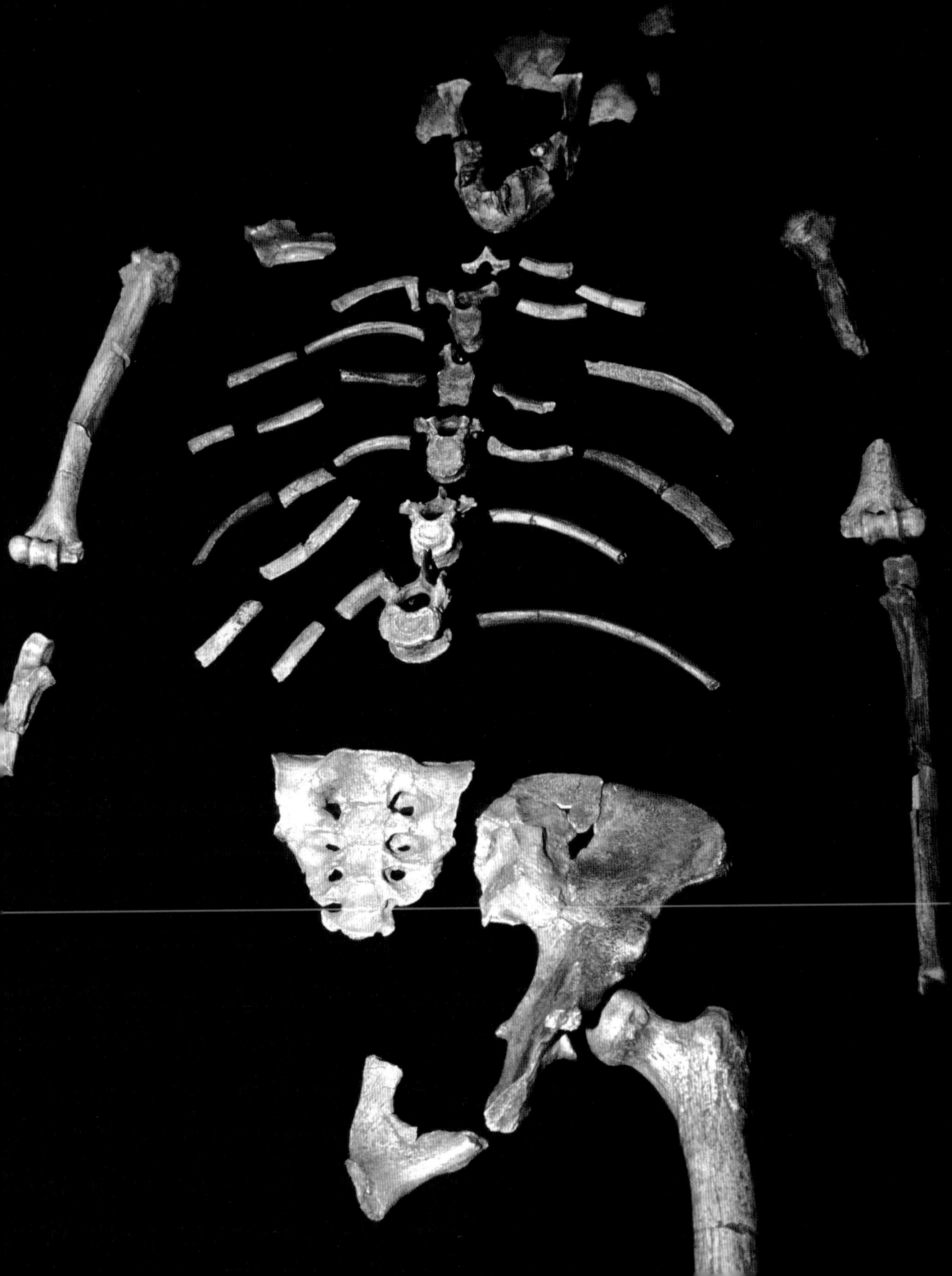

TIMELINE

1891 | 1912 | 1924 | 1927 | 1949 | 1953 | 1960

1891 *Homo erectus* (in the form of Java Man) is discovered by Eugène Dubois.

1912 The fossil remains of what was claimed as an early human are discovered in Piltdown, UK, and named the Piltdown Man.

1924 Taung Child is discovered in South Africa, the first specimen of *A. afarensis*, but his importance is disputed.

1927 A collection of bones is found in a cave in China, later named Peking Man.

1949 Remains of *Homo ergaster* are discovered in Africa, the earliest known human species, but they are not recognized as such at the time.

1953 Piltdown Man is revealed as a fake, made of parts less than 50,000 years old.

1960 In Africa, Jonathan Leakey discovers fossil remains of *Homo habilis*, a new species of man and the earliest traced to Africa.

1859
CREATED OR EVOLVED?

When Darwin published *On the Origin of Species by Means of Natural Selection* in 1859, it was immediately contentious. Darwin was suggesting that nature changes over time, and that humans, too, could have changed over time. It was taken by many as an affront to the Biblical account of Earth's history. In some places, that view is still widely held. For other people, it suggested new avenues for science to explore. One of those people was Eugène Dubois, who went first to Sumatra and then to Java to hunt for an early human ancestor. In 1891, he found one: the first early hominid discovered, *Homo erectus*, who lived around a million years ago.

1924–1960
LOOKING FOR THE MISSING LINK

When the Taung Child was found in Taung, South Africa in 1924, it was first dismissed as an unimportant ape. It took many years for the importance of this, the first instance of *A. afarensis*, to be recognized. Instead, attention focused on Peking Man, found in China in 1927, which was consistent with the belief that humans evolved in Asia or Europe. We now know that *Homo erectus* evolved from *Homo ergaster*, who migrated from Africa around 1.8 million years ago. *Homo ergaster* was discovered in 1949, but not identified until 1975. Africa was finally confirmed as the cradle of humanity in 1960 when *Homo habilis* was found in Africa. At 1.75 million years old, it was older than any *Homo* species then recognized.

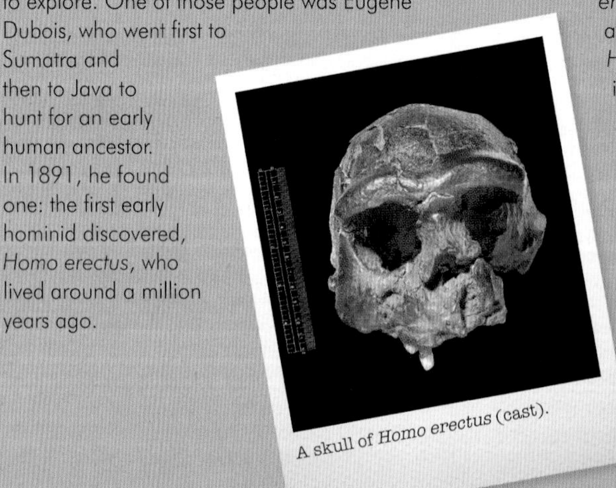

A skull of Homo erectus (cast).

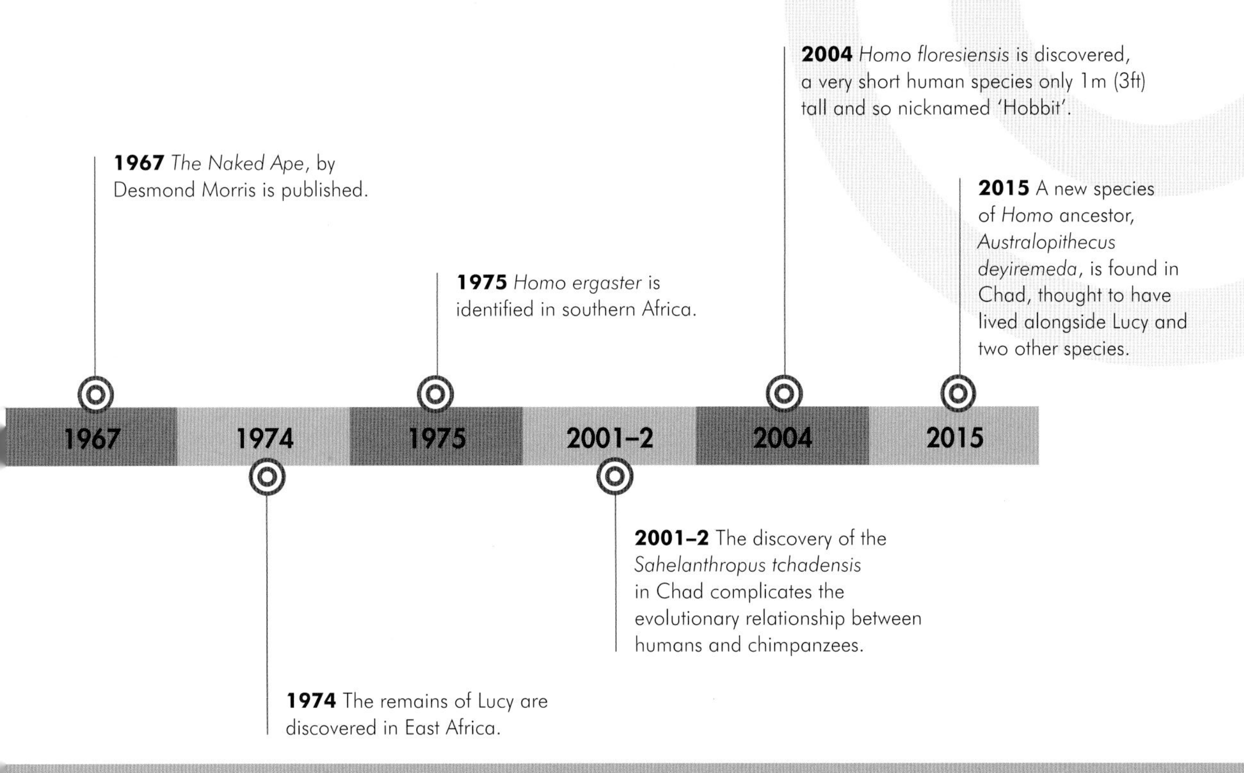

1967
AN APE AFTER ALL

Despite the research linking modern humans with earlier hominids, and the ongoing search for a common ancestor who marked the point at which our evolution parted company with that of chimpanzees, few people really thought of humans as animals like any others. Their complacency was challenged by a book called *The Naked Ape*, published by Desmond Morris in 1967. Morris was the keeper of mammals at London Zoo and presented humans for the first time in the same way as other mammals are studied – with descriptions of habit, mating behaviour, rearing of young, fighting and so on. Humanity's specialness was stripped away completely.

2001
A TANGLED FAMILY TREE

At the time of Lucy's discovery, only around eight human ancestors were known; around 20 are now identified. It is no longer assumed that there was a direct line of descent from a common human/chimpanzee ancestor to modern humans. There might have been dead ends, with early human species dying off or breeding with others and not necessarily leading to modern humans in a straight line. In 2001 and 2002, the discovery of Toumai, *Sahelanthropus tchadensis*, in north Africa added further complications. At 6–7 million years old, Tumai could be a common ancestor of chimpanzees and humans – or be related to both but an ancestor of neither.

The skull of Toumai (cast).

1975

ALTAIR 8800: THE DAWN OF PERSONAL COMPUTING

'There is no reason for any individual to have a computer in his home.'

– Ken Olson, founder of Digital Equipment Corporation, 1977

An Altair 8800 computer.

Personal computers, and their progeny in the form of smartphones, tablets and other 'devices', are a feature of everyday life. Yet it is little more than 40 years since the very first personal computer hit the shops. Or, actually, the mail-order listings in hobby magazines. It could be bought pre-assembled, or as a build-it-yourself kit. Once it was built, the owner had to program it. And reprogram it every time it was turned on, as it didn't have any means of storing programs or data. Nor did it have an easy way of entering or examining them, as there was no keyboard, screen or mouse.

The first personal computer was the Altair 8800, launched in 1975. It was programmed by flicking switches and showed its results by flashing lights. It could be built in any box the enthusiast had handy. Despite these limitations, it was hugely popular. Ed Roberts, who designed and made the Altair kits, expected to sell only a few hundred but in fact sold several thousand in the first month after it was featured in the January issue of *Popular Electronics*. The first programming language for the Altair was Altair BASIC – the first product of the new company Micro-Soft (it had a hyphen in those days). Both the idea of a personal computer, and Microsoft, took off.

It was not a smooth ride from the Altair to the present, though. Initially, a host of small computer companies set up business, many starting in garages in California, as Apple did. Each of these came up with its own architecture for hardware and software and the result was myriad incompatible systems. The move towards conformity began only when IBM introduced its own personal computer in 1981. Over the coming years, other systems were slowly squeezed out of the market, leaving only Apple with its own proprietary system and the industry-standard IBM-PC running MS DOS. Although MS DOS came to be used on a large number of PC-clones, Apple kept the rights to its operating system for its own computers – a distinction it still maintains.

FLASHPOINT FACT

Bill Gates and Paul Allen claimed that Micro-Soft was already developing a programming language that could run on the Altair 8800 and arranged a meeting to demonstrate. In fact, they had nothing – not even an Altair. Paul Allen finished writing it on the plane and the first time it ever ran on an Altair was at the demonstration.

TIMELINE

1642 Blaise Pascal invents the Pascaline, a mechanical calculator.

1801 The jacquard loom is the first machine to be controlled by punched cards with coded instructions.

1820s Charles Babbage designs but does not build his Difference Engine and Analytical Engine, two mechanical programmable computers,

1936–8 Konrad Zuse builds an electro-mechanical programmable computer, the Z1.

1943 Colossus, built secretly at Bletchley Park in England, is the first programmable electronic computer.

1948 'Baby', the first stored-program computer, is built in Manchester, England.

1951 Maurice Wilkes develops 'microprogramming' to control different aspects of the computer itself, effectively separating the operating system from other programs.

1975 The Altair 8800 goes on sale, the first-ever personal computer.

1642	1801	1820s	1936–8	1943	1948	1951	1975

FLASHPOINTS

1642
TOWARDS A COMPUTER

The idea that a machine could take over some of our mental work is old. Blaise Pascal built the 'Pascaline' mechanical calculator in 1642, prompted by Pascal helping his father with burdensome calculations as a supervisor of taxes. In 1801, the jacquard loom stored complex weaving patterns on punched cards – the first use of 'programming', though the machine was entirely mechanical. Around 20 years later, Charles Babbage put calculating and program together, at least in theory. He designed two machines in the 1820s that followed programmed instructions, written for Babbage by Ava Lovelace, but did not have the funds to build them.

Blaise Pascal with his 'Pascaline'.

1943
COLOSSUS AND OTHER COLOSSI

The claim to being the first programmable computer actually built is hotly contested. Konrad Zuse built the Z1 in 1936–8, though it and its construction plans were destroyed in the Second World War by Allied air raids on Berlin in 1943. It was a partially programmable calculator. The war also saw the development of Colossus at Bletchley Park in England, built to crack German secret codes in 1943. The work was top secret, and the machines and plans for them were destroyed after the war. With everyone who worked on it sworn to secrecy for decades, Colossus was generally overshadowed by the later American military computer, ENIAC, built in 1946.

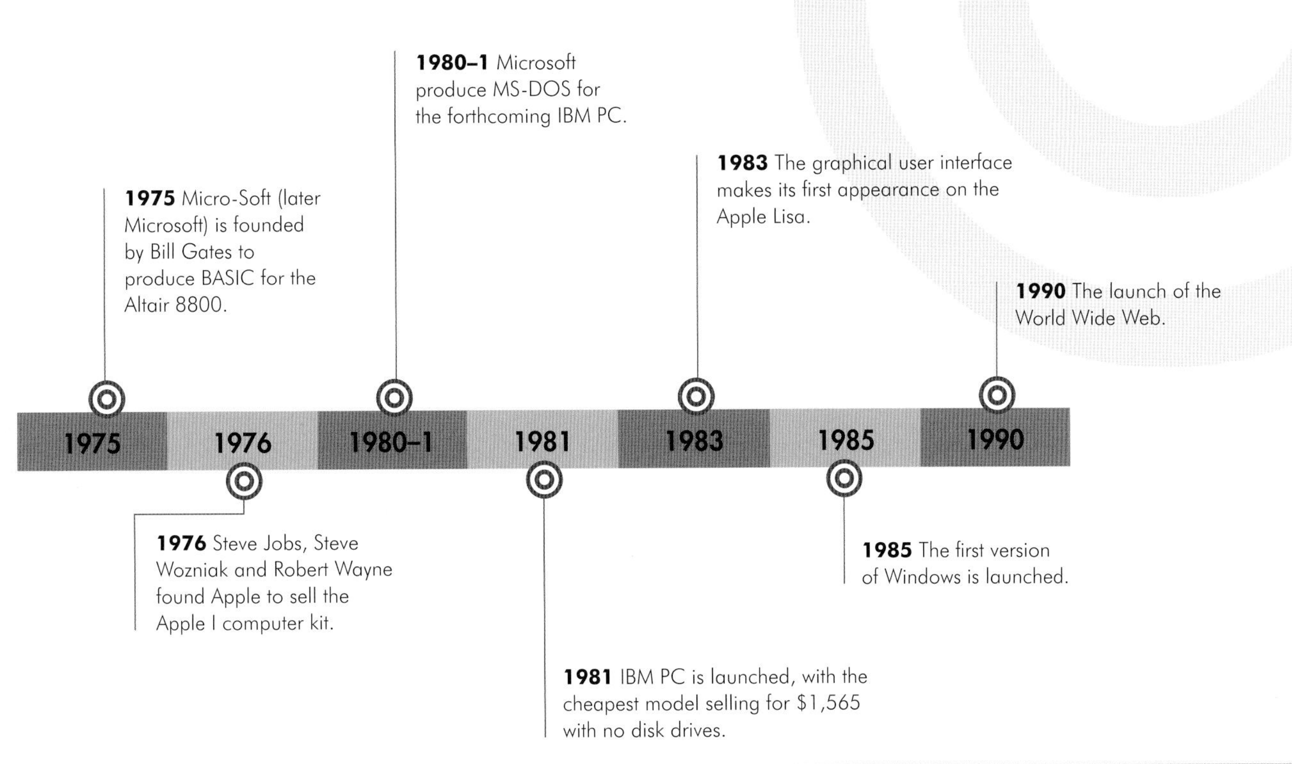

1980–1
PERSONAL COMPUTING COMES OF AGE

The kit computers such as the Altair 8800 appealed to enthusiasts, but as soon as manufacturers began to produce ready-built computers with a keyboard and monitor, that would do useful things, personal computers took off in the mainstream. Initially, there was no conformity – one computer was not compatible with another, and they all used different operating systems and required different programs. By the end of the 1980s, though, this had resolved into the dominance of the IBM-PC and Apple, with other manufacturers largely switching to PC-compatible products in order to survive

1983
WINDOWS ONTO THE WORLD

The next big change was the development of the graphical user interface (GUI), the familiar desktop with windows, icons and documents, navigated using a mouse and menus. The original concept was designed at Xerox Park, as an experimental idea for use on large computers, but was soon copied (Xerox claimed stolen) by Apple. The Apple Lisa, released in 1983, was the first computer to use a GUI. The interface was used for other Apple computers, and emulated in the form of Windows for PC-compatible computers in 1985. The new interface made computers easier to use – no one needed to remember strings of apparently meaningless commands – and their appeal rapidly widened to even the least technically sophisticated of home users.

The Apple Lisa.

1977

EXPLORING THE SOLAR SYSTEM – AND BEYOND

'Mars has been flown by, orbited, smacked into, radar examined and rocketed onto, as well as bounced upon, rolled over, shovelled, drilled into, baked and even blasted. Still to come: Mars being stepped on.'

– Buzz Aldrin, 2013

The rings of Saturn, photographed by the *Voyager* mission in 1982.

From the start of the 1960s, the USA and the USSR were racing to the moon, but they were also both beginning to explore the wider solar system. The most ambitious project was the launch of *Voyagers 1* and *2* in 1977. They were set to fly close to all four of the solar system's largest planets – Jupiter, Saturn, Uranus and Neptune – which were about to line up in a way that happens only once in every 176 years. But, beyond that, they would carry on travelling, leaving the solar system and heading out into space and towards other stars.

Now, in the second decade of the 21st century, *Voyager 1* has gone further from Earth than any other man-made object. After photographing Jupiter and Saturn, it set a course for interstellar space and is now beyond the heliosheath – the outer limit of the sun's area of influence. *Voyager 2* photographed all four large planets and is a little behind *Voyager 1*. Both *Voyagers* carry a 'Golden Record', a message for any extraterrestrials that might encounter the craft. It contains a library of sounds and greetings from Earth and of images of Earth and human artifacts.

Some people have worried that the Golden Record could be seen as advertising a planet rich in resources that might attract hostile attention; a few even disapprove of the images of naked humans it shows.

The *Voyagers'* plutonium power supplies have a half-life of 88 years and might run out between 2020 and 2025. Although this will eventually prevent some critical systems from working, and will finally end communication with the *Voyagers*, the craft will not stop moving. They will continue forever until found or destroyed. On its current trajectory, *Voyager 1* will pass within 1.7 light years of another star in 40,000 years. Some of the original *Voyager* team suggested that just before *Voyager* runs out of battery power it should be sent an instruction to use its final fuel to change direction and propel it into a solar system, maximizing the slim chances of it being found around 60,000 years in the future.

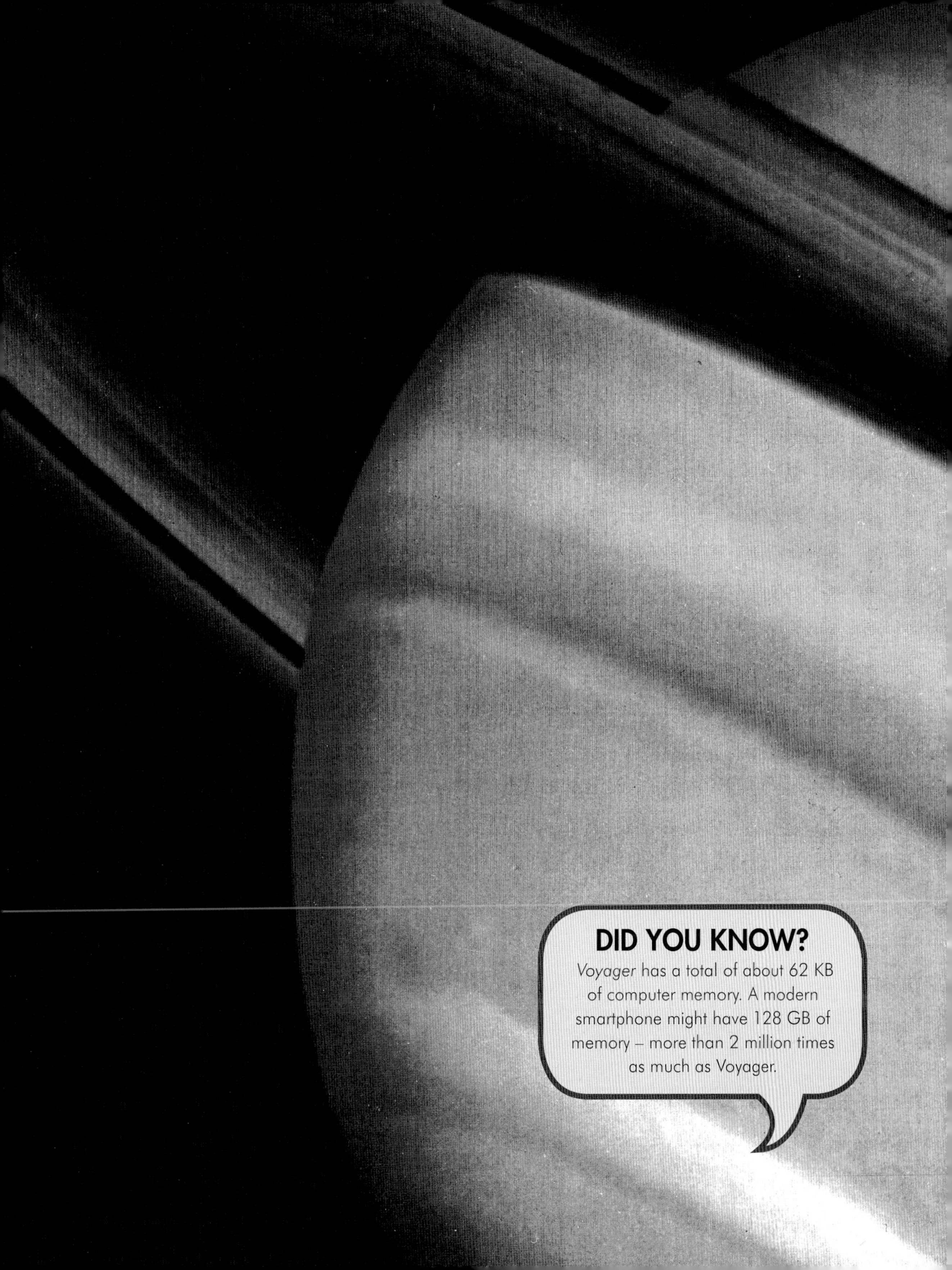

DID YOU KNOW?

Voyager has a total of about 62 KB of computer memory. A modern smartphone might have 128 GB of memory – more than 2 million times as much as Voyager.

TIMELINE

1961 The Soviet *Sputnik 7* makes the first attempt to fly to Venus, but a rocket failure dooms the mission.

1962 NASA's *Messenger 2* makes the first successful flyby of Venus at 34,000km (20,740 miles), discovering ground temperatures up to 428°C (802°F).

1965 *Venera 3* (USSR) is the first object from Earth to impact with another planet; no data is returned as Venus's atmosphere crushes *Venera*.

1967 *Venera 4* returns data on the temperature and pressure of Venus's atmosphere.

1971 *Mars 3* (USSR) becomes the first spacecraft to land (not crash) on Mars.

1972 *Pioneer 10* (USA) is the first spacecraft to enter the asteroid belt.

1978 *Pioneer Venus 1* maps the whole surface of Venus, sending photos back to Earth.

1982 *Venera 13* returns the first colour photographs from Venus.

1989 *Magellan* (USA) maps 98 per cent of the surface of Venus to a resolution of 100m (328ft).

FLASHPOINTS

1970
EARLY EXPLORATION

In the early 1960s, both the USSR and then the USA sent probes – unmanned spacecraft – to Venus. The earliest craft had a low rate of success, with few even getting to the planet. The first to land, the Soviet *Venera 3*, returned no data before it was crushed by the Venusian atmosphere. The first to return data from the surface of another planet was *Venera 7* in 1970. It reported surface temperatures of 475°C (887°F) and an atmosphere of almost pure carbon dioxide at pressures of 90 times the Earth's atmospheric pressure. *Venera 9* visited in 1975 and sent back the first photographs from the surface of another planet.

1971
TO MARS

Mars has always been an attractive to planet to explore, being the most Earth-like and often our nearest neighbour (depending on orbits). But until recently, success has been limited. The USSR launched nine missions to Mars between 1960 and 1969, which all failed. In 1971, *Mars 2* crashed on Mars and *Mars 3* achieved a controlled landing but stopped transmission after 14 seconds. *Vikings 1* and *2*, launched by NASA in 1975, became the first successful missions to Mars, with landers collecting samples and sending back colour photographs of the planet and orbiters mapping the planet. Subsequent Mars trips have left rovers that have returned data over a long period. A priority is to discover whether there has ever been life on Mars.

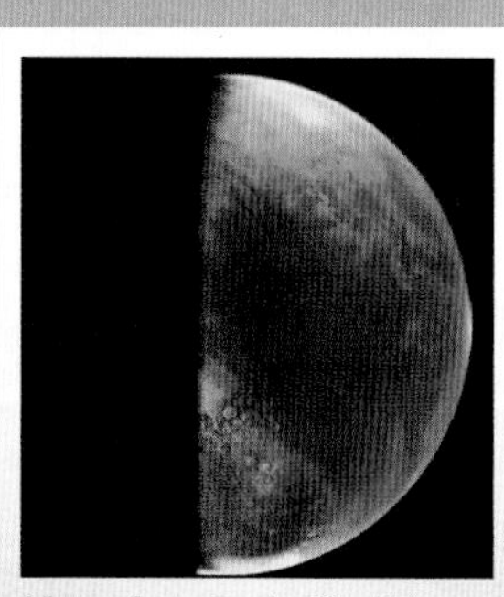

Mars, photographed by the *Viking 1* orbiter.

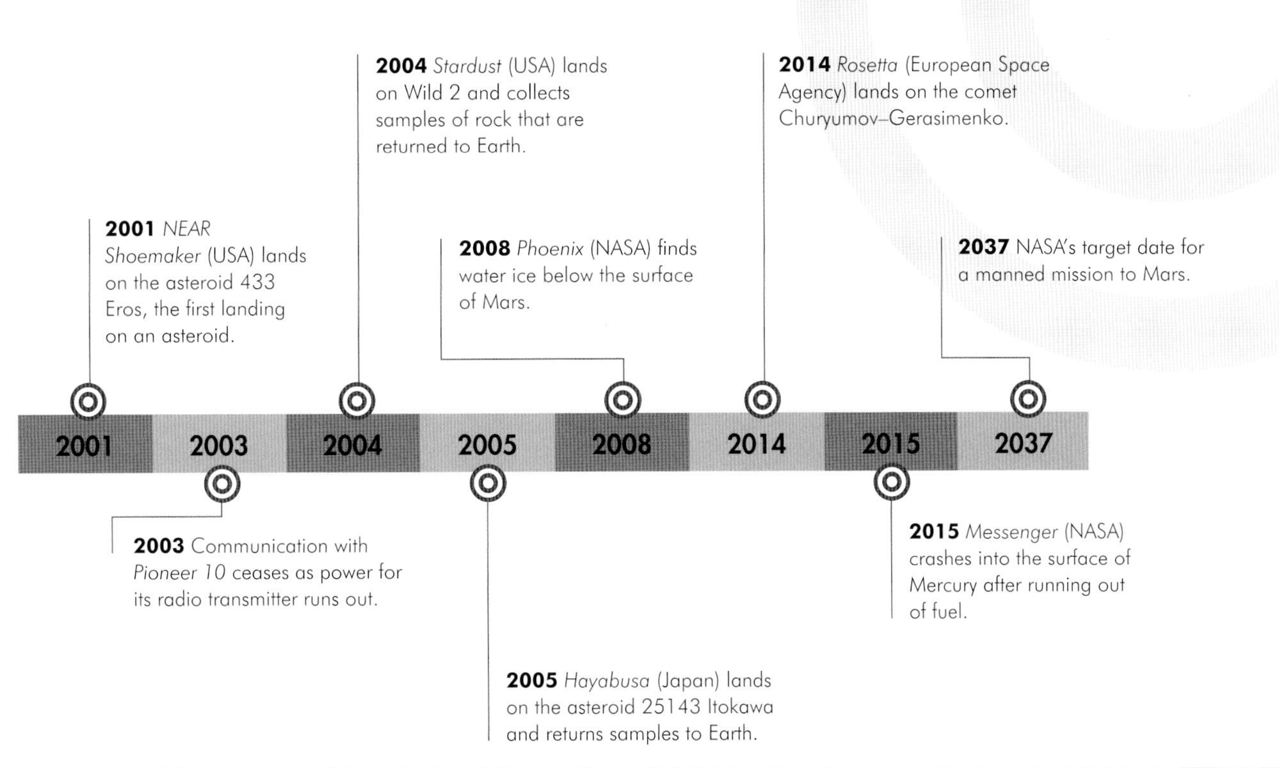

1972
PIONEERS

The *Pioneer 10* mission (NASA), launched on 3 March 1972, photographed Jupiter and its moons before heading off towards interstellar space. It ceased communication with Earth in 2003. It is now more than 100 AU (astronomical units) from the Sun, but still 68 light years from the next star on its course – it will take two million years to reach it. *Pioneer 11*, its sister craft, was the first probe to photograph Saturn and investigate the asteroid belt. Both *Pioneers* carry a plaque on the outside that shows a picture of naked humans, and pictures that aim to show the location of Earth.

Jupiter.

1989
ASTEROIDS AND COMETS

Since the mid-1980s, spacecraft have visited a number of comets, asteroids and dwarf planets, either on flyby missions or with landers. *Galileo* was launched in 1989 to investigate and photograph Jupiter and its moons. It also became the first craft to flyby an asteroid, 951 Gaspra, and to discover an asteroid moon, Dactyl, which orbits 243 Ida. In 2001, the NASA craft *NEAR Shoemaker* became the first ever to land on an asteroid, 433 Eros. In 2004, *Stardust* landed on Wild 2 and not only collected samples of rock but, for the first time, returned them to Earth.

1978

IVF: THE FIRST TEST-TUBE BABY

'The greatest of all curses is the curse of sterility.'

– Franklin D. Roosevelt, 1910

Robert Edwards, holding Louise Brown with Patrick Steptoe (right).

IVF – *in vitro* fertilization – has brought joy to millions of parents since the first 'test-tube baby' was born in 1978. The technique involves fertilizing a human egg cell outside the body and then implanting it to grow normally in the uterus. It was developed to help couples with infertility problems. It is now sometimes used to screen embryos for life-limiting or other serious genetic conditions if the parents are known to be carriers. IVF is also now used with other animals in farming, conservation work and reproductive research.

After the first successful *in vitro* fertilization of a human egg in 1969, it was nearly ten years before the first human IVF pregnancy. The technique was perfected in rabbits and hamsters and only attempted in humans once it was clear that the offspring developed normally to adulthood and had a normal lifespan for their species. Louise Brown was the first baby born after IVF treatment, on 25 July 1978 in Manchester, England. The process and pregnancy were overseen by Patrick Steptoe and Robert Edwards. They removed an egg through a tiny incision in the mother's abdomen. The egg was fertilized with the father's sperm in liquid that mimics conditions in the woman's body. After growing in a petri dish (*in vitro*) for two and a half days, the egg was returned to the mother's body for a normal pregnancy. The petri dish used for Louise Brown is still preserved in the Cambridge fertility clinic where her life began.

Steptoe and Edwards faced hostility and suspicion when they began to experiment with human IVF. The Medical Research Council in Britain refused them funding. They set up their project independently, and had no difficulty finding infertile couples happy to be part of their research – though participants were sworn to secrecy for their own protection. The process was a strain for parents, and had a success rate of only 12 per cent in its early days. By 2013, 35 years after the birth of Louise Brown, five million babies had been born by IVF. The procedure has become far more streamlined and straightforward, and the success rate is now around 25 per cent.

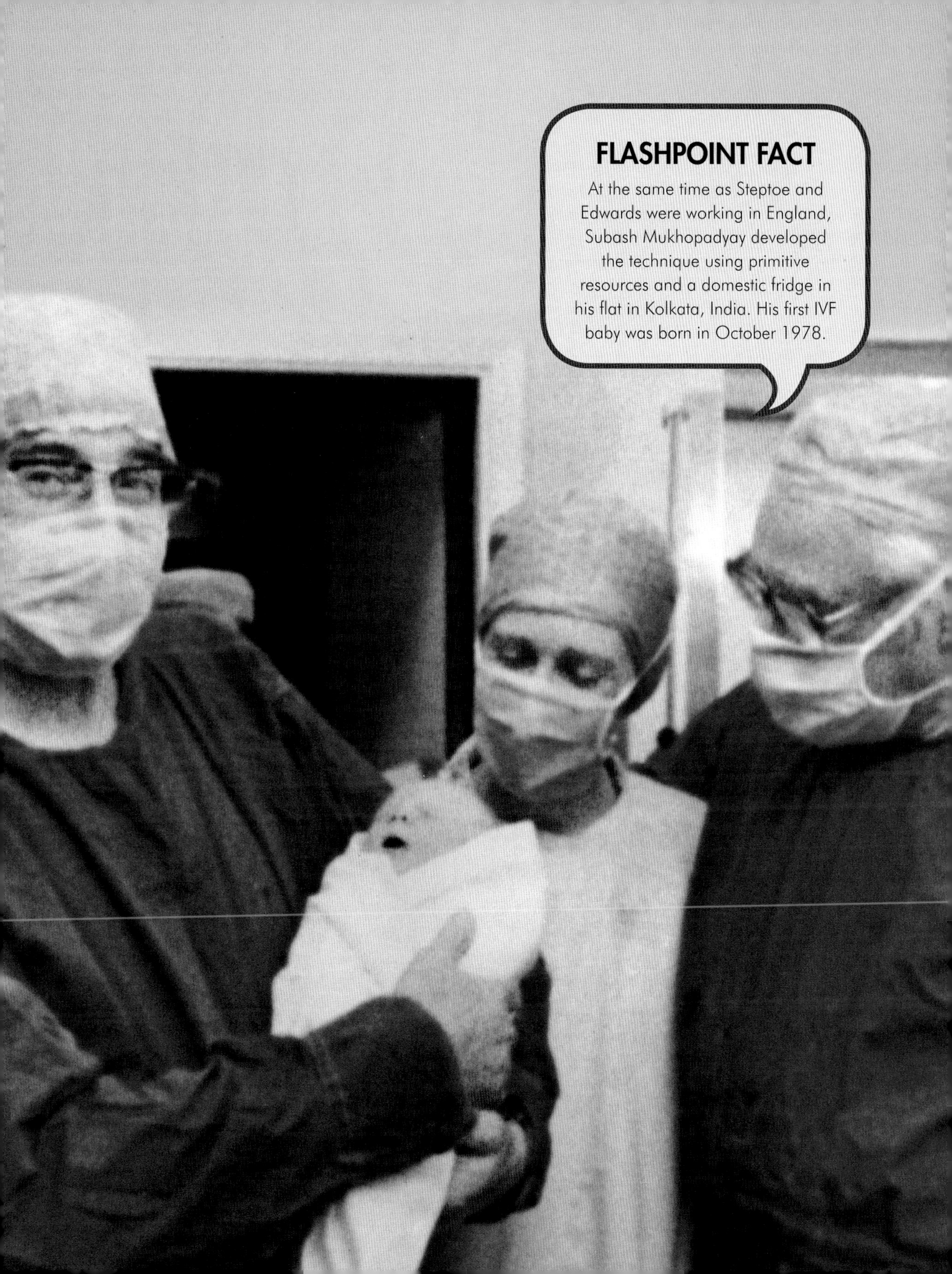

FLASHPOINT FACT
At the same time as Steptoe and Edwards were working in England, Subash Mukhopadyay developed the technique using primitive resources and a domestic fridge in his flat in Kolkata, India. His first IVF baby was born in October 1978.

TIMELINE

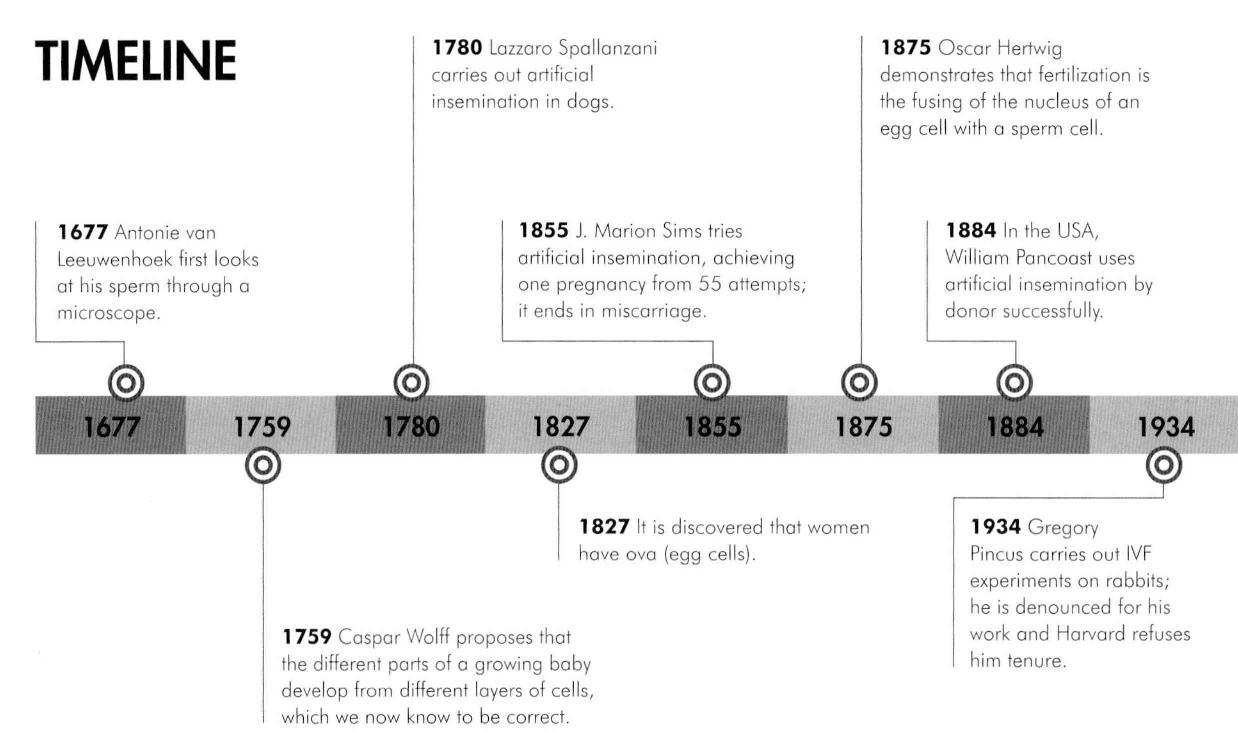

1780
EGG + SPERM = BABY

Until the 18th century, people believed that all the important material for a baby was carried in the father's sperm, the mother providing only an environment in which the baby grows. The Italian biologist Lazzaro Spallanzani showed in 1780 that both egg and sperm are needed to create offspring, a result he found by carrying out artificial insemination in dogs. The earliest-known human artificial insemination by donor was unethical by any standards. The woman was under general anaesthetic and unaware of William Pancoast's plan. Her husband was infertile and the couple had been unable to conceive. Pancoast used donor sperm to inseminate the woman, who gave birth nine months later. Pancoast later told the woman's husband what he had done.

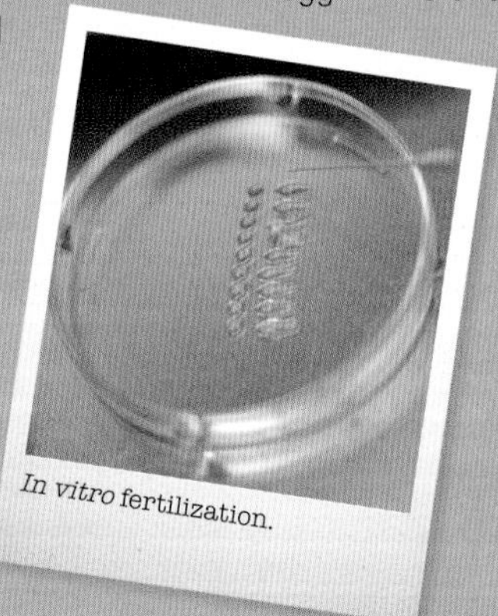
In vitro fertilization.

1934
STARTING WITH IVF

Early experiments in IVF met with popular and professional hostility, which made progress slow. In 1934, Gregory Pincus succeeded in fertilizing rabbit eggs *in vitro*, but when he reported his findings his work was criticized and Harvard refused to give him tenure. It was another ten years before John Rock and Miriam Menkin successfully fertilized a human egg *in vitro* after leaving the egg and sperm in contact for longer than usual. They made no attempt to implant the egg. Even so, in 1949 the Vatican denounced all IVF work.

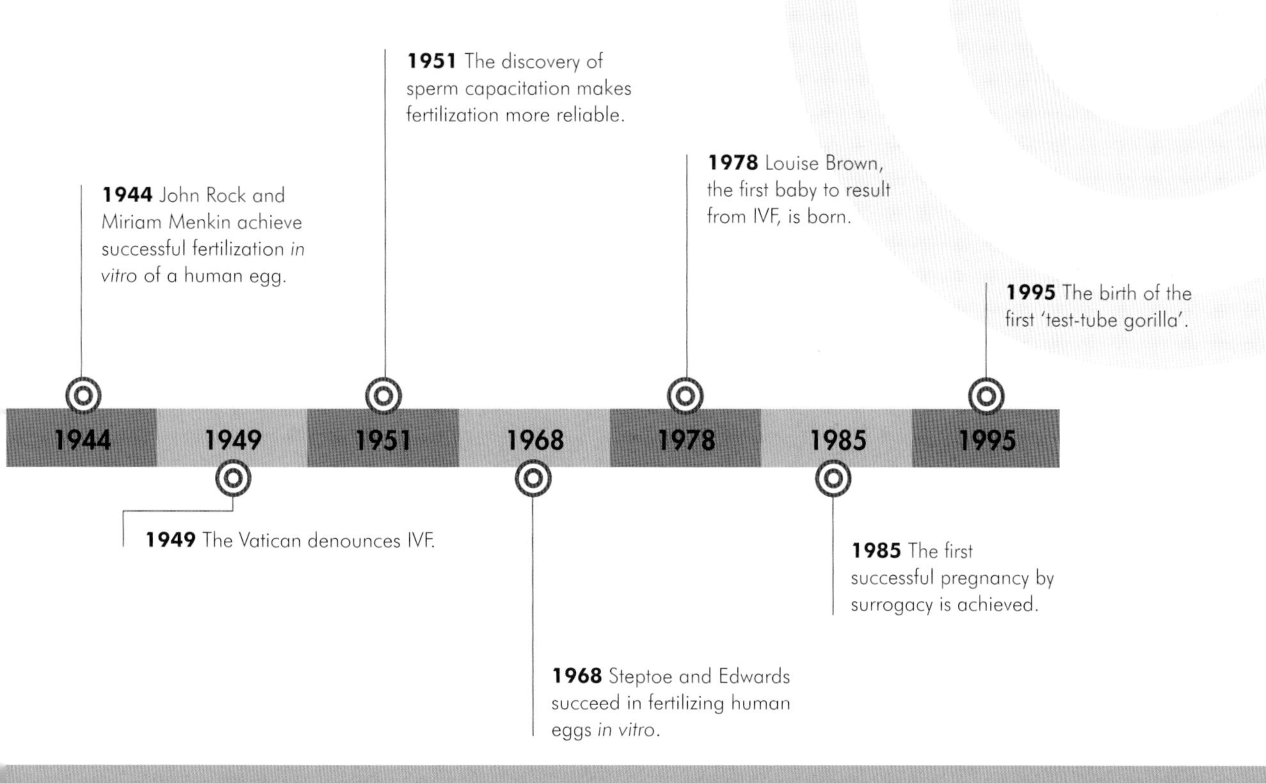

1951

WORKING WITH SPERM

The most important breakthrough came in 1951 with the discovery of sperm capacitation – chemical changes that take place in the sperm inside the woman's body before the egg is fertilized. This stage thins the membrane around the sperm, making it easier for it to fuse with an egg. Triggering capacitation by exposing the sperm to suitable chemicals and the right temperature made *in vitro* fertilization much more reliable. In 1961, Daniele Petrucci in Italy claimed to have grown a human embryo for 29 days, by which time it had developed a heartbeat. He then destroyed it. The following year, he claimed to have successfully implanted a fertilized egg in a patient, but this was never confirmed.

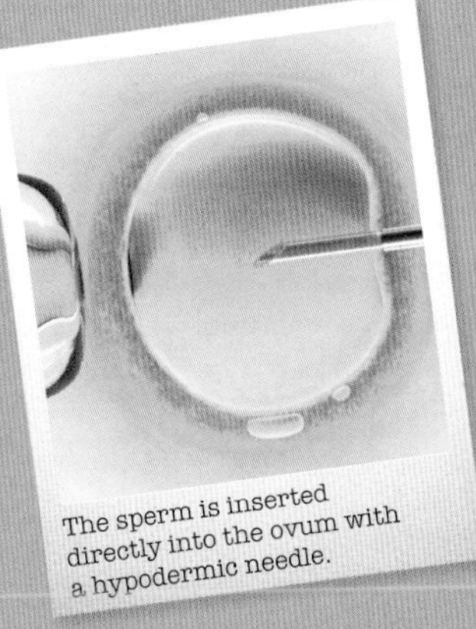

The sperm is inserted directly into the ovum with a hypodermic needle.

Post 1978

BOUNCING BABIES

Since Louise Brown, more than a million babies have been born after IVF. The technique has been used, sometimes controversially, to enable post-menopausal women to have babies, and, from 1985, with surrogate mothers who grow the child for genetic parents who are not able to go through a pregnancy. It is used alongside genetic testing, too, since it allows a batch of eggs to be fertilized, then any carrying a serious genetic condition to be rejected, implanting only healthy eggs back into the mother. IVF and artificial insemination have also both been used to help increase the numbers of endangered species.

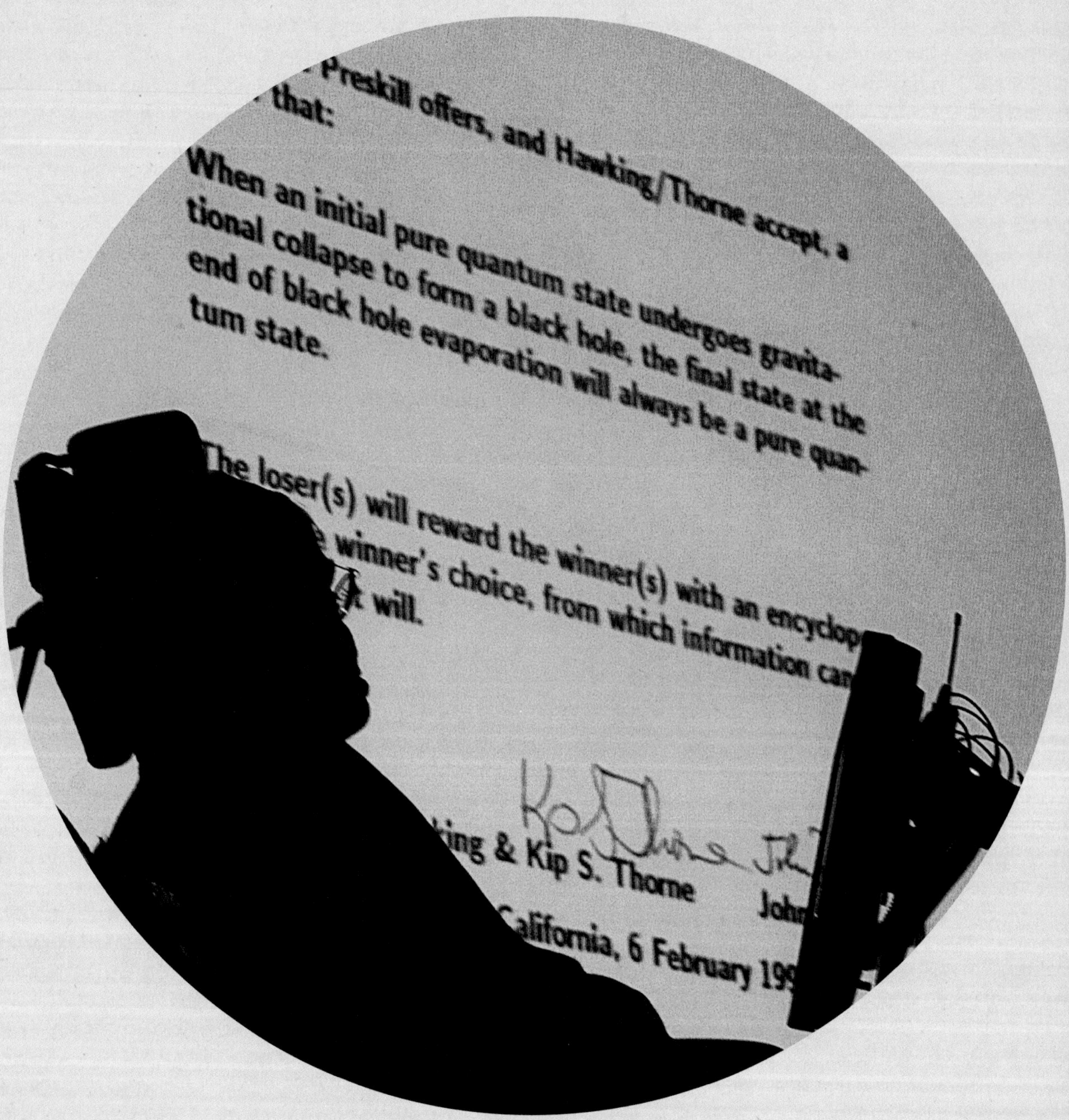

Preskill offers, and Hawking/Thorne accept, a
that:

When an initial pure quantum state undergoes gravitational collapse to form a black hole, the final state at the
end of black hole evaporation will always be a pure quantum state.

The loser(s) will reward the winner(s) with an encyclop
winner's choice, from which information can
will.

king & Kip S. Thorne
John
California, 6 February 19

1980–1999

The final decades of the 20th century saw some of the promise of early discoveries coming to fruition. The unravelling of DNA led to myriad new developments, ranging from genetic engineering through cloning and stem-cell therapies. In some regards, the last two decades were another period of optimism and ambition. Again it looked as though we had answered many of the questions posed by science. Of enormous help in solving those problems was the development of powerful computers, which doubled in processing power every two years. Calculations that would have taken centuries without computers are now feasible. They have contributed not only to high-powered scientific applications, but made possible computer-animated films and the dominance of the World Wide Web since the 1990s.

There has been, though, a strange repetition of the pattern of the late 19th century. Just as it looked as though many of science's great puzzles were being solved, new challenges emerged. The great problems that face the scientists of the 21st century were already emerging in the 1990s. They include the question of how to tackle climate change and its consequences and – less pressingly – what the universe is made of. It seems that all we thought we knew accounts for only a tiny proportion of what there is.

There are plenty of opportunities still for the next generation of scientists, and there will be many more flashpoints in the 21st century, some of which we can only just begin to imagine. Others would no doubt be as alien and astonishing to us if we could look forwards to 2100 as the World Wide Web, the eradication of smallpox and the existence of quarks would have been to someone peering forward in 1900.

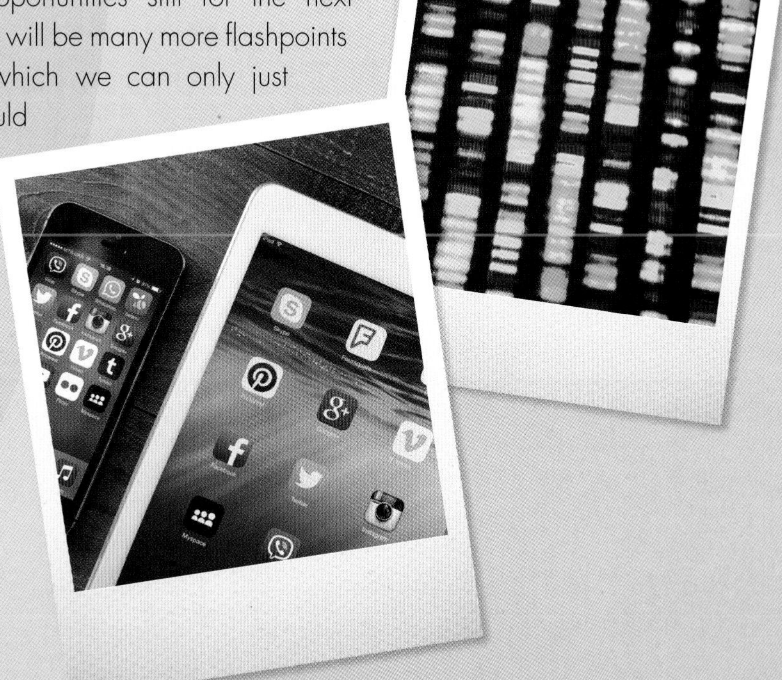

1980

DEATH TO SMALLPOX

'The world and its peoples have won freedom from smallpox, which was a most devastating disease sweeping in epidemic form through many countries since earliest time, leaving death, blindness and disfigurement in its wake.'

– WHO declaration, 1980

One of the last smallpox cases, in 1975.

Smallpox has been one of the most terrible diseases humanity has ever faced. It existed for at least 12,000 years and killed 300–500 million people in the 20th century alone. Its eradication in 1980 was one of the greatest scientific achievements of all time.

Smallpox is characterized by a rash of small pustules that covers the entire body and kills around 30 per cent of sufferers. Protective inoculation began in the late 18th century, but the modern campaign to eradicate the disease began in 1958, when the World Health Organization (WHO) stated a resolution to end smallpox. In 1967, this became the Smallpox Eradication Programme. This was made possible by an improved vaccination procedure with a two-pronged needle. It could be used by non-specialist staff, was quicker and used less vaccine than the 'gun injectors' used previously. The programme had two main activities: 100 per cent vaccination rate in all susceptible populations, and rigorous monitoring to trace, report, quarantine and treat all cases. WHO workers travelled around with a picture of a baby affected by smallpox to help people identify the disease, and paid a reward to anyone who reported a case. When a sufferer was identified, they and their contacts were vaccinated and quarantined. By following its programme aggressively in all countries, the WHO succeeded in wiping out smallpox in the wild in 1978. In 1980 the disease was classified as officially eliminated.

FLASHPOINT FACT

In 1978, after smallpox had been eradicated in the wild, medical photographer Janet Parker died from smallpox contracted in a laboratory in the UK.

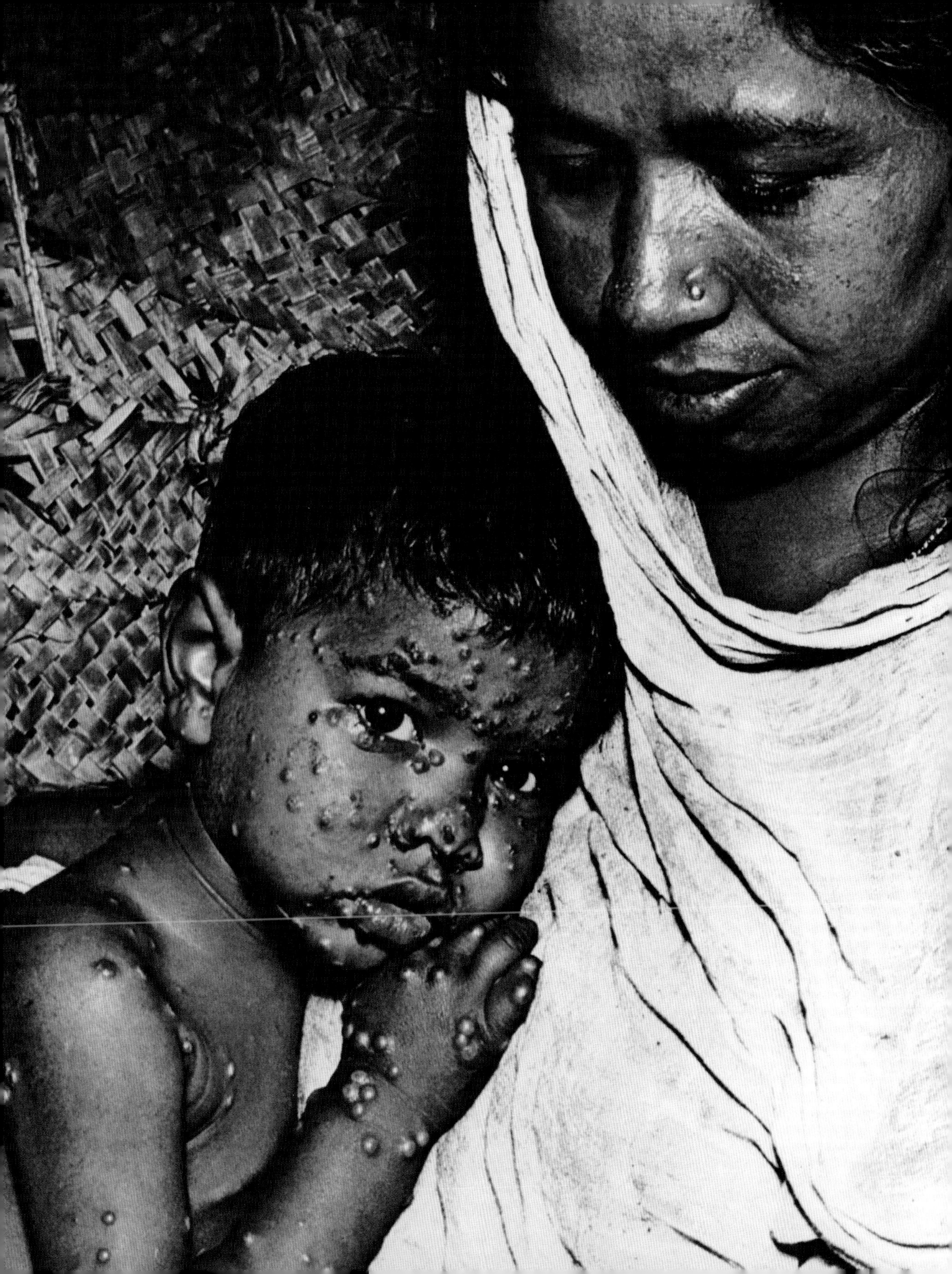

TIMELINE

c 1721 | 1774 | 1796 | 1853 | 1949 | 1965

c. 1721 Lady Mary Wortley Montagu introduces variolation (an early form of vaccination via exposure) to Europe from Turkey.

1774 English farmer Benjamin Jesty immunizes his family using cowpox, but does not publicize it or perform tests or more immunizations.

1796 Edward Jenner makes an effective vaccine against smallpox using matter from cowpox pustules.

1853 Vaccination against smallpox becomes compulsory in the UK.

1949 Freeze-drying of a vaccine is achieved, meaning it can be safely transported long distances even in warm climates.

1965 The introduction of the bifurcated needle means less vaccine is needed for each patient.

FLASHPOINTS

93BC
AN OLD DISEASE

The earliest firm evidence of smallpox is in Egyptian mummies 3,000 years old. In Asia, a preventive measure called variolation was practised at least 1,000 years ago. It involved exposing patients either to dried matter or fluid from smallpox pustules. The patient developed some symptoms but they (usually) recovered. Lady Mary Wortley Montagu encountered variolation in Turkey and introduced it to Europe in 1713.

In 1796, the English doctor Edward Jenner developed a vaccine made from matter from cowpox pustules. Cowpox is a related but less dangerous disease often contracted by milkmaids. Learning that milkmaids never catch smallpox, Jenner developed his vaccine. It was effective, and safer than variolation.

1830s and 1840s
COMPULSORY VACCINATION

Following initial resistance and ridicule, Jenner and his vaccine became increasingly popular as people saw that it worked. The Royal Society refused to publish his findings, so he had them privately printed in a book that became a bestseller. The native tribes of North and South America were particularly badly hit as they had no immunity to smallpox. In 1832, the USA federal government introduced a vaccination programme for Native Americans. In 1853, the British government introduced a law requiring everyone to be vaccinated. Various states in the USA adopted compulsory vaccination from 1844 onwards.

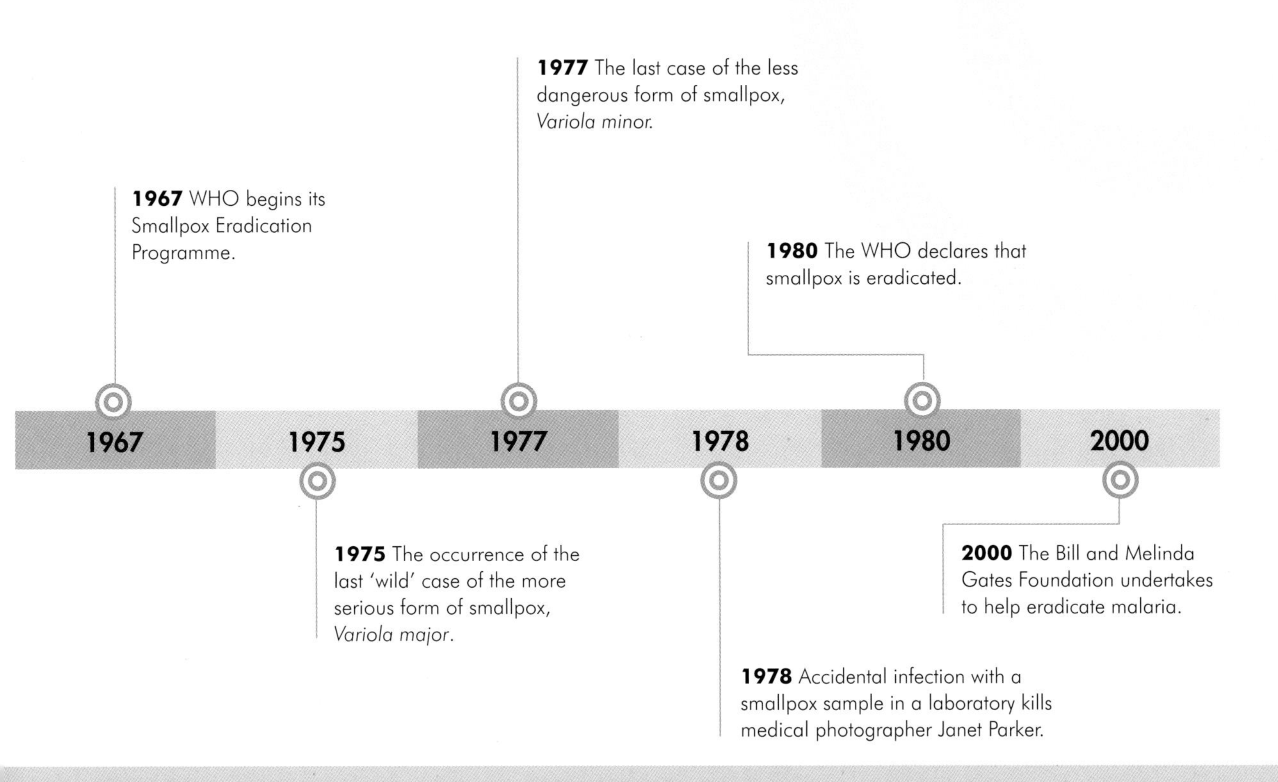

1900
FIGHTING FOR OUR LIVES

By the start of the 20th century, smallpox had become mainly a disease of the poor and of less developed countries. When an epidemic started in New York in 1900, it was assumed that poor and black people would be affected most and so little was done to contain it. Against this backdrop of ignorance, smallpox killed an estimated 300–500 million people during the first 80 years of the 20th century. The World Health Organization decided in 1958 to target the disease for eradication, as there was already an effective and safe vaccine. The invention of the bifurcated needle in 1965 meant that less vaccine was needed for each inoculation.

Post 1980
GONE BUT NOT FORGOTTEN

Although smallpox no longer exists in the wild, samples are kept in two research laboratories. One of these is in Atlanta, USA and the other in Novosibirsk, Russia. These are kept as reference samples and could be used to make a vaccine again if necessary. There is some controversy surrounding the retention of the virus, with some people voicing worries that it could escape, be stolen or be released maliciously or weaponized. The threat of weaponized smallpox, perhaps in the hands of terrorists, is also an argument *for* keeping the samples. As *Variola* is a simple virus, it could be constructed from scratch by terrorists with the right equipment, so the potential need for a vaccine is real.

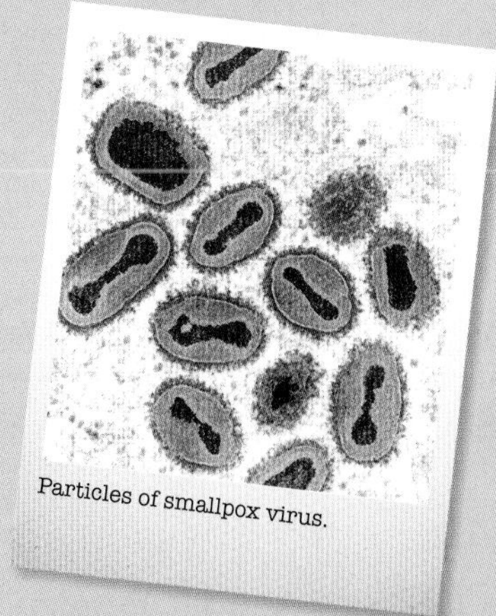
Particles of smallpox virus.

1981

IT ALL STEMS FROM STEM CELLS

'Stem-cell research is the key to developing cures for degenerative conditions like Parkinson's and motor neurone disease from which I and many others suffer. The fact that the cells may come from embryos is not an objection, because the embryos are going to die anyway.'

– Stephen Hawking

Human embryonic stem cells.

When an animal starts to grow from a fertilized egg, it is initially a single cell, then two identical cells, then four, and so on as the cells divide. At some point, the cells have to differentiate – become different types. Pluripotent stem cells – those that can still develop into cells of many kinds – were first discovered in 1963.

Embryonic stem cells were first isolated and cultivated outside the body in 1981. The techniques established by Martin Evans and Matthew Kaufman enabled them to grow a line of stem cells in a petri dish, make genetic modifications to the stem cells, then reintroduce them into a female mouse to grow into a genetically modified mouse. The modified mice mature normally and can reproduce. This process is now used to create the transgenic mice used in medical research into human diseases.

The huge potential for stem cells relies on scientists being able to work out how to trigger them to grow into a particular type of tissue. Therapeutic use would then lie in prompting the regeneration of, say, heart muscle in a patient with heart disease, and nerve cells of the appropriate type in a patient with Parkinson's or Alzheimer's.

One benefit offered by stem cells is that if a patient's own stem cells can be used to grow replacement tissue there will be no risk of rejection – their immune system will recognize them as a legitimate part of the body. Stem cells can also be used to test new drugs (therefore removing the initial need to test on animals or humans) and to produce new tissue for victims of burns.

FLASHPOINT FACT

Sea urchins have been used in many observations and investigations of early embryo development and cell differentiation; their eggs are transparent, so everything can easily be seen.

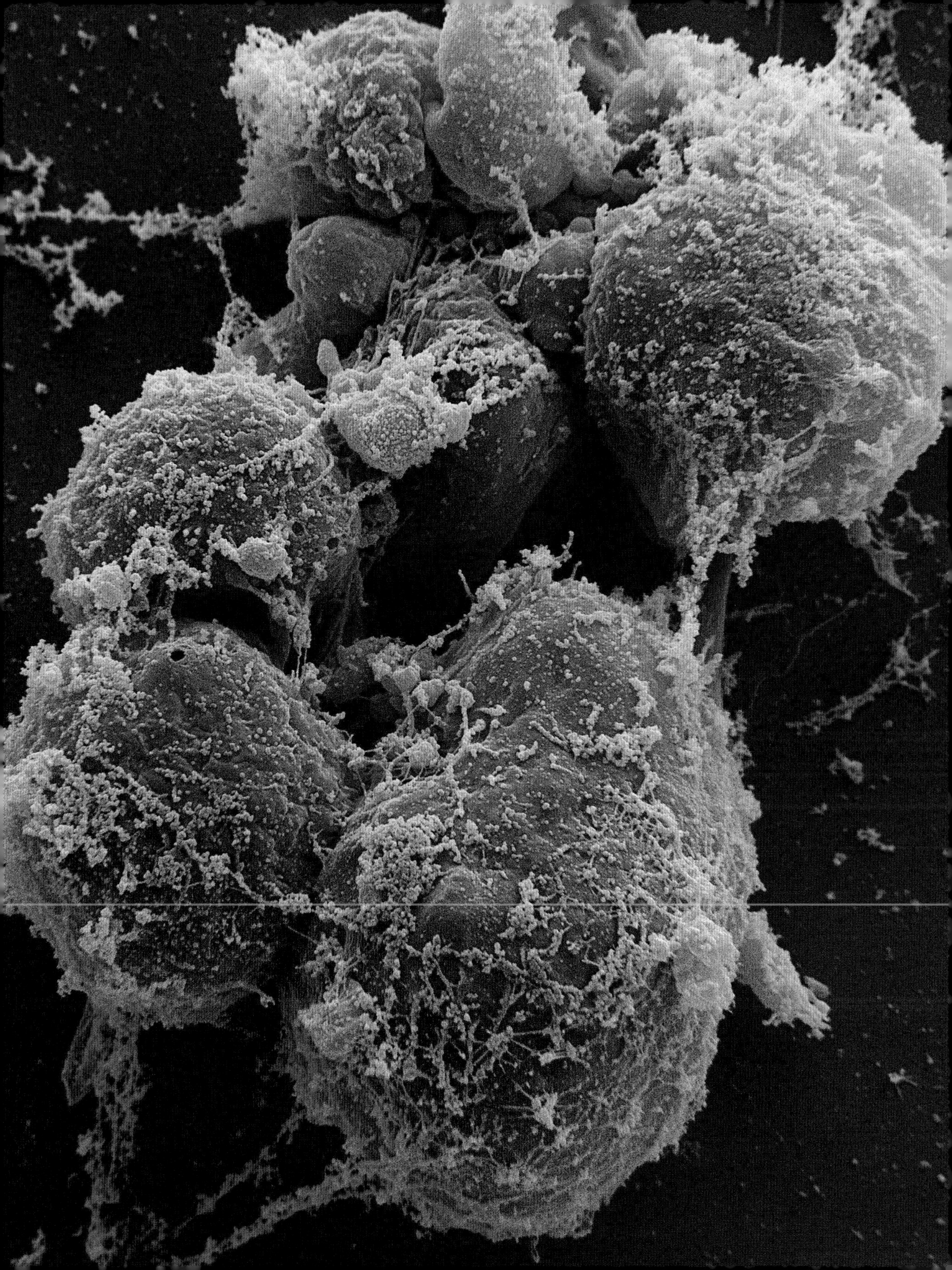

TIMELINE

1907	1909	1963	1968	1978	1981

1907 Max Askanazy finds embryonic cells capable of producing tumours composed of different types of tissue (teratomas) if injected into rats.

1909 Alexander Maximor announces that all blood cells come from the same cellular 'ancestor' (now called haematopoietic stem cells).

1963 Ernest McCulloch and James Till discover that different types of blood cells grow from a special type of cell found in bone marrow.

1968 E. Donnall Thomas carries out the first successful bone-marrow transplant, using bone marrow from the patient's sister.

1978 Haematopoietic stem cells are found in the blood of the umbilical cord.

1981 Martin Evans and Gail Martin isolate embryonic stem cells from mice and cultivate them.

FLASHPOINTS

1907
TERRIBLE TUMOURS

The first research into stem cells took place in Germany in the second half of the 19th century. Several scientists were interested in how cells, originally all the same in an early embryo, come to differentiate and produce all the tissues and organs of the body. In 1907, Max Askanazy described an experiment in which he had taken cells from an early rat embryo and injected them into the abdominal cavities of adult rats. These grew into teratomas, or 'monster tumours' – a type of tumour that contains mixed tissues, often including flesh, hair and even teeth. From this he concluded that the rat stem cells had the potential to change into any of the tissue types required to make a rat.

Embryonic stem cells.

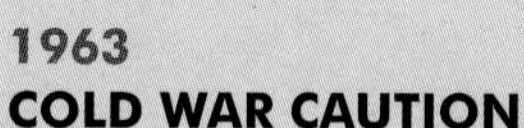

1963
COLD WAR CAUTION

At the height of the Cold War in the 1960s, Canadian researchers Ernest McCulloch and James Till were investigating the effects of massive doses of radiation on mice in anticipation of a possible nuclear war. Transplanting bone marrow between healthy and irradiated mice in 1963, they discovered that nodules appeared in the spleens of the mice. These were producing blood cells of all three types: red blood cells, white blood cells and platelets. They had found a type of cell that could differentiate into any of the blood cells – haematopoietic stem cells. The first successful bone-marrow transplant therapy for leukaemia followed in 1968.

1989 The first knock-out mouse is created, using modified stem cells to disable a gene.

1997 Dominique Bonnet and John Dick show that leukaemia is produced by the same stem cells that produce normal blood cells – cancer consists of cells that have gone off course.

1998 Two teams at different US universities simultaneously grow lines of human stem cells in the laboratory.

2001 Funding for stem-cell research with new lines of human cells is stopped in the USA.

2006 Shinya Yamanaka and Kazutoshi Takahashi create induced pluripotent stem cells from adult cells (in rats).

2013 A 3-D printer is produced at Herriot-Watt University in the UK that can print groups of human stem cells.

1989	1997	1998	2001	2006	2013

2001
ETHICAL ISSUES

Because stem cells can grow into many different types of tissue, they seem to offer considerable promise for medicine. What if damaged nerve or liver cells could just be regrown from stem cells? In practice, it has turned out to be quite difficult to use stem cells in this way and progress is rather slow. In addition, research has been stalled by ethical debate. In the US, funding for new lines of human stem cells was suspended between 2001 and 2009. Many people object to the destruction of human embryos – which could otherwise by implanted in a woman's uterus and grow into a baby. This has fuelled the search for methods of reverting adult cells to pluripotency.

Research using human embryonic stem-cell research is controversial and has many opponents.

2006
CELLS GOING BACKWARDS

In 2006, Shinya Yamanaka in Kyoto, Japan, found a way of inducing pluripotency in adult cells – of returning differentiated cells to their pluripotent state so that they could once again act as stem cells. The technique involved inserting genetic material into the cell, using a carrier virus. In 2014, Haruko Obokata and Charles Vacanti claimed to have found a simple technique of returning cells to pluripotency by exposing them to acid. Other researchers were unable to reproduce their results, though, which were widely denounced as either flawed or faked.

1988

STEPHEN HAWKING AND THE TRUTH ABOUT BLACK HOLES

'We are each free to believe what we want and it is my view that the simplest explanation is there is no God. No one created the universe and no one directs our fate. This leads me to a profound realization. There is probably no heaven, and no afterlife either. We have this one life to appreciate the grand design of the universe, and for that, I am extremely grateful.'

– Stephen Hawking

Stephen Hawking delivers a lecture in Berkeley, California, in 1988.

Stephen Hawking is surely the most famous living physicist. Diagnosed with the degenerative neural condition motor neurone disease at the age of just 21, his entire career has been conducted in the grip of increasing disability.

As a research student at Cambridge, Hawking became interested in the work of Roger Penrose on the origins of the universe, the fate of stars and the creation of black holes. Hawking's own work came to focus on black holes. The discovery that made him famous within the scientific community came in 1974, when he revealed that black holes are not quite as all-consuming as they seem. In fact, he said, radiation should be able to escape from the edge of a black hole. This hypothetical radiation became known as Hawking radiation. It provides a way for small black holes (smaller than the mass of the moon) to shrink and eventually evaporate.

Hawking became famous outside scientific circles with the publication of *A Brief History of Time* in 1988. This book tries to explain 20th-century cosmological theory in a form that lay readers can understand. A later version, *A Briefer History of Time*, was easier to understand.

In 2014 the award-winning film *The Theory of Everything* popularized Stephen's story and sparked a fresh wave of public interest in his works and his life.

FLASHPOINT FACT

Hawking had a long-running bet that the Higgs boson would never be found. When it was found at CERN in 2013, he quickly admitted defeat and said Higgs should win the Nobel Prize (which he did).

IMAGINERY
UNIVERSITY OF

TIMELINE

1915 Albert Einstein publishes the General Theory of Relativity.

1916 Karl Schwarzchild uses Einstein's theory of relativity to describe black holes and defines their gravitational radius (Schwarzchild radius).

1931 Georges Lemaître publishes his theory that the universe began from a singularity, later called the Big Bang theory.

1942 Stephen Hawking is born in Oxford, England.

1963 Hawking is diagnosed with motor neurone disease.

FLASHPOINTS

1916
BLACK HOLES

The phenomena later known as black holes were first predicted in 1916 by Karl Schwarzchild. When a star at least three times the mass of the sun dies, it leaves behind a core that collapses in on itself under its own gravity. It becomes super dense – such a star would occupy a sphere about the radius of New York city. A black hole grows by drawing in other matter. It can be very destructive, ripping matter away from passing stars or engulfing entire planets or stars. The new matter is also hugely compressed. At the centre of a black hole is a 'singularity' where density and gravity are infinite.

Bright bursts of gamma rays indicate the creation of a black hole by a star collapsing.

1974
HAWKING RADIATION

The common view of a black hole is that nothing can escape its extremely powerful gravitational pull. Hawking proposed in 1974 that at the very edge of a black hole – the event horizon, the point where that pull becomes irresistible – particle–anti-particle pairs are created. Their creation uses energy from the black hole's gravity. One of the pair is sucked into the black hole and the other escapes. It should be observed as black-body radiation. The particle that has escaped was produced using energy from the black hole and so represents a net loss of energy (and so mass) to the black hole. If a black hole is not increasing in mass by drawing in other matter, it will slowly shrink and eventually evaporate in this way.

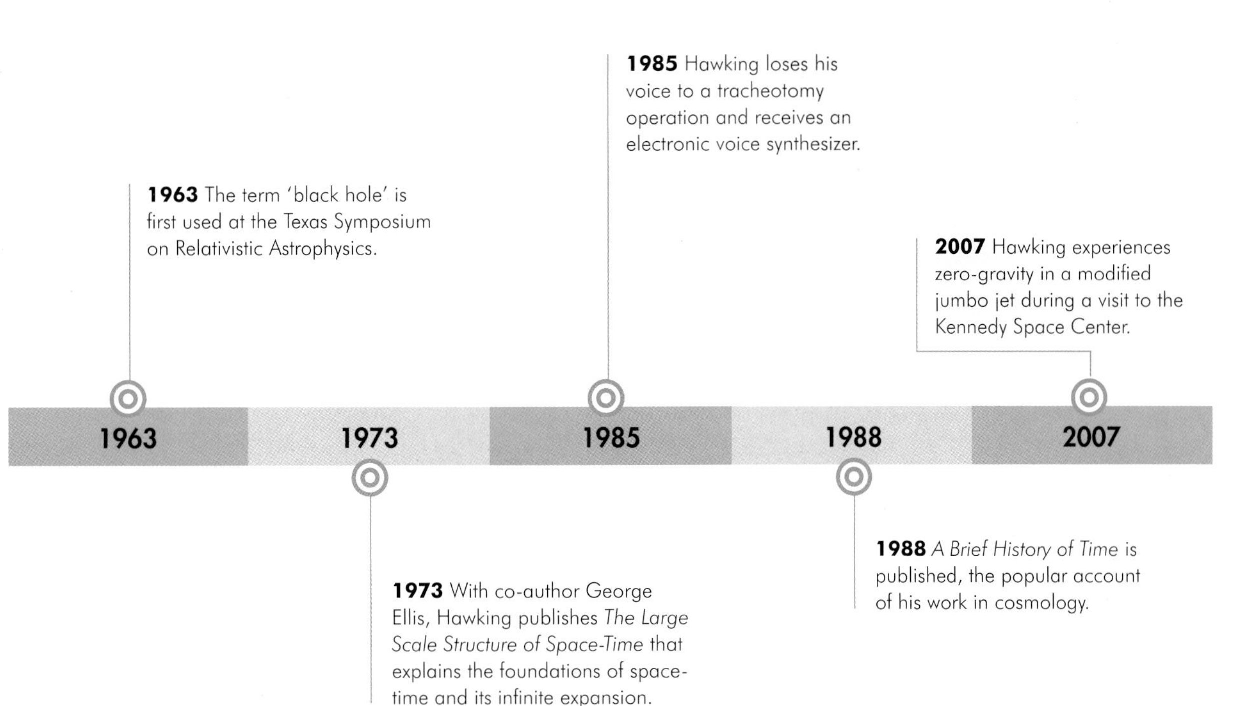

1981
SPACE WITHOUT BOUNDARIES

In 1981, Hawking suggested that there might be no boundary to the universe – meaning that there was no beginning or end to it as well as no edge to it in space. This makes 'before the Big Bang' a meaningless concept; time did not exist before the Big Bang. This is rather like saying that if you stand at the North Pole, the concept 'north' has no meaning as there is nowhere further north to go. Hawking has also suggested that the universe could have many histories, which add together to give its present state. In his model, the traditional singularity from which the universe erupted is replaced with something more like the North Pole.

Post 2015
THE FUTURE

Hawking is pessimistic about the prospects for humanity. He has several times voiced his opinion that humankind is set on a course of self-destruction, whether through war, climate change, the release of a genetically engineered virus or some other means. He has suggested that although the development of artificial intelligence is possible, it should be avoided: the outcome of creating beings that will become more intelligent than ourselves is very uncertain, and it could be both the biggest and last event in human history. He has promoted space exploration, believing that we might need to expand to other planets in order to survive. He says that it is highly likely that aliens exist, but that trying to contact them would be inadvisable.

Some of the oldest galaxies in the universe, photographed by the Hubble Space Telescope.

1990
THE HUMAN GENOME PROJECT

'I believe that reading our blueprints, cataloguing our own instruction book, will be judged by history as more significant than even splitting the atom or going to the moon.'

– Francis Collins, director of the National Human Genome Research Institute

A DNA sequence. The coloured bands indicate base pairs in the structure of the DNA.

By the mid-1960s, it was clear that DNA is the genetic material and that it carries the 'recipe' for an organism in the form of codons (triplets of bases) that each map to an amino acid. Putting the amino acids together in the sequence in which they occur in the DNA of a gene builds the correct protein. This is the essence of the mechanism of inheritance. In order to put it to good use, though, we need to know what each gene is and what it does. This was the challenge faced by the Human Genome Project, an international research programme begun in 1990. The team, led by James Watson, decided not to sequence the genome of an individual but to mix samples from a number of anonymous donors.

There are 3.3 billion base pairs in the human genome, making up 32 chromosomes divided into around 20,500 genes, though that number was not known at the start of the project. It took ten years to get the first draft of the sequence, which shows a list of all the pairs of bases in the order in which they occur. The draft had to be published quickly because a private firm, Celera Genomics Corporation, had joined the race and intended to patent important parts of the genome for profit. The final version of the full sequence was published online in 2003 and is free for anyone to use

But that's not the end of the story. Although the entire sequence of base pairs has been listed, scientists know the location and function of only some of the genes. There are many still to identify, along with the ways in which pairs or groups of genes might interact.

FLASHPOINT FACT

Humans share 98.8 per cent of their DNA with chimpanzees and around 44 per cent with fruit flies.

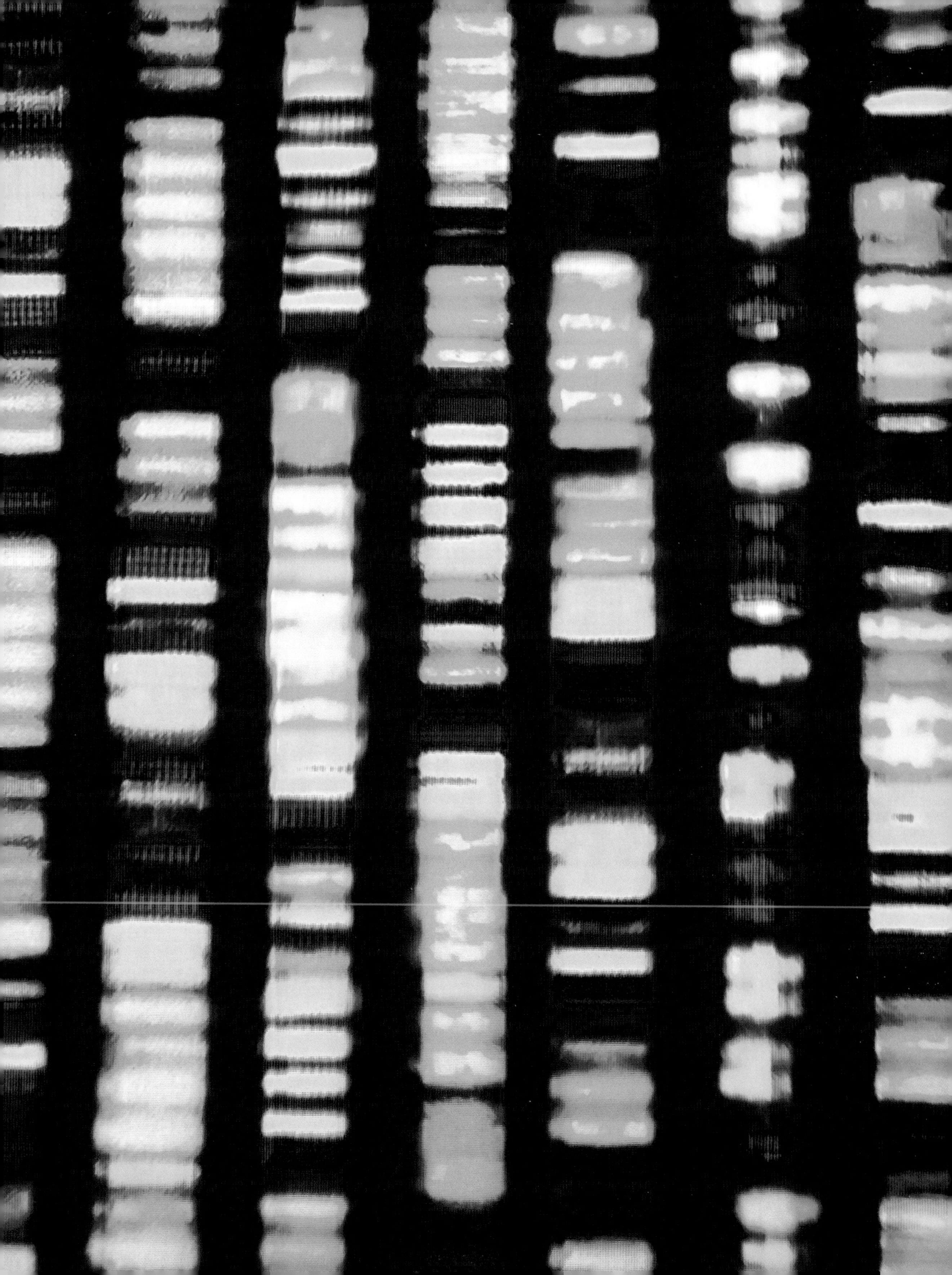

TIMELINE

1964 Charles Yanofsky and Sydney Brenner demonstrate that one gene does indeed code for one protein.

1969 The first gene is isolated at Harvard; it is from the bacterium *E. coli*.

1977 Fred Sanger publishes the first genome to be sequenced, that of the virus ΦX174.

1977 Fred Sanger develops a technique for rapidly sequencing DNA.

1983 Kary Mullis invents the polymerase chain reaction (PCR), which hugely speeds up genetic sequencing.

1983 Mapping of the first human genetic disease, Huntington's chorea, is achieved.

1989 A gene mutation that causes cystic fibrosis is discovered.

1995 The first bacterial genome is sequenced, *Haemophilus influenza*.

FLASHPOINTS

1964
FINDING THE GENES

Soon after Marshall Nirenberg discovered in 1961 that the sequence of bases in codons codes for a specific amino acid, came confirmation of the one gene, one protein theory. In 1964, Charles Yanofsky and Sydney Brenner discovered through their work with *E. coli* that mutations in a particular gene cause a defect in a corresponding protein. By changing areas of the gene, they found they could alter the sequence of amino acids in part of the protein built. This means that if we know which gene codes for each protein – that is, we know the purpose of each gene – we are well equipped to make changes to genes directly, perhaps to fix hereditary disorders.

1977
SEQUENCING

Initially, sequencing of DNA – working out the order of base pairs – was a very laborious process that could deal with only a few hundred base pairs every day. In 1977, Fred Sanger and his team completed the sequencing of the first organism – a bacteriophage (a virus that affects bacteria) called X174. It has only 5,368 base pairs on a single strand of DNA, but sequencing it was still a formidable achievement. The same year, Sanger published his newly developed method of rapid sequencing, in which a lot of the work was automated.

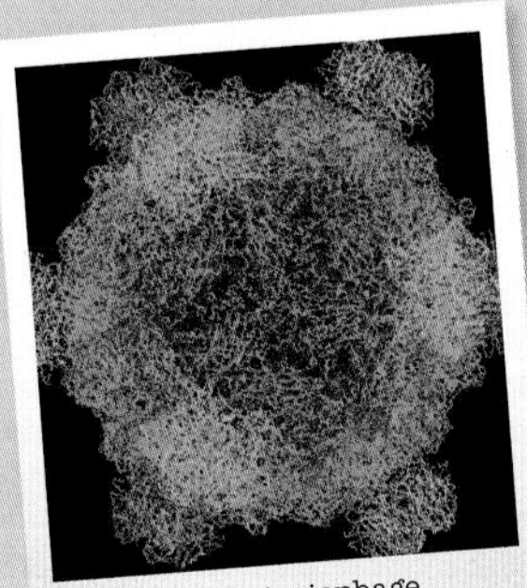

A model of bacteriophage ΦX174.

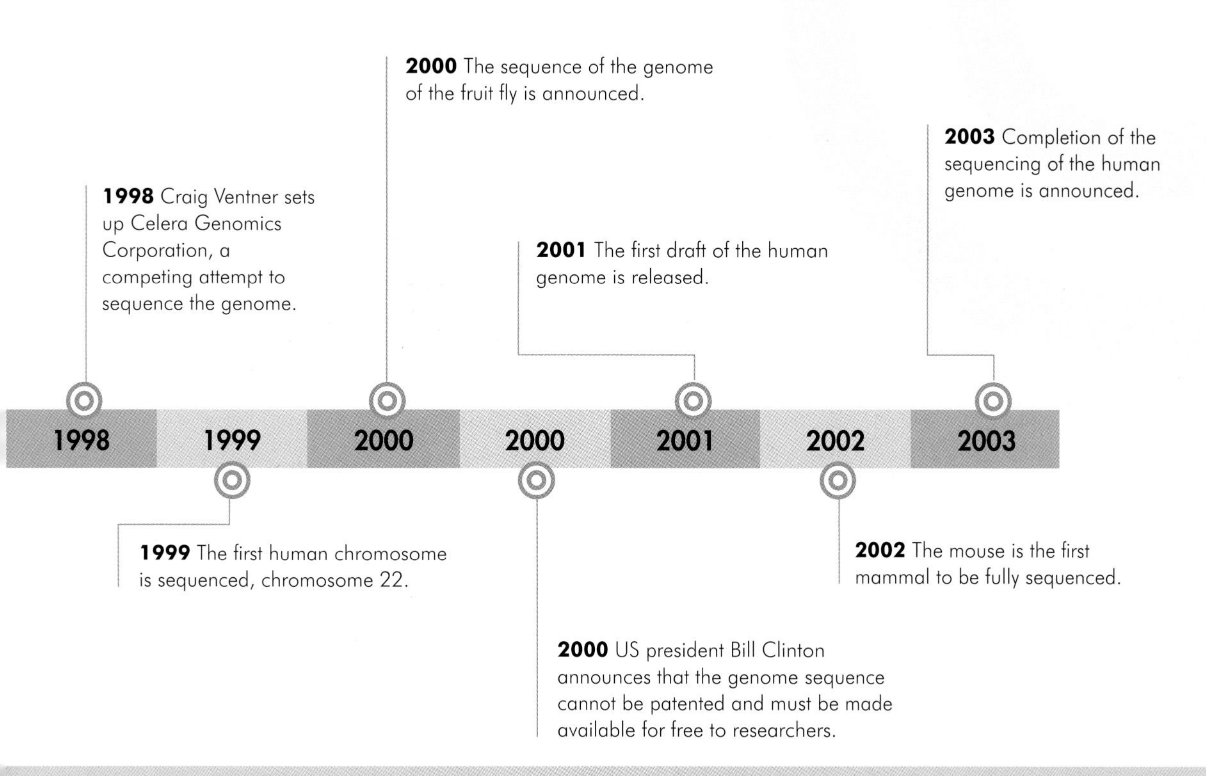

1983
GENETIC DISEASE

The first human genetic disease to be mapped was Huntington's chorea. This causes neurons in the brain to die, leading first to jerky movements, then to dementia and finally to death. Onset is usually in midlife. In 1983, it was discovered that mutation of the HD gene on chromosome four is responsible for Huntington's. The mutation that causes the disease adds increasing repetitions of the base sequence 'CAG', which codes for the protein huntingtin. Although this knowledge does not yet help us cure Huntington's, it does offer a test for it. The test can be carried out prenatally, to screen for the disease while termination of the pregnancy is possible, or at any stage after birth.

2005
USES OF THE GENOME

The most immediate uses of the genome sequence are medical: to identify the causes of genetic disorders and to screen or test for them. At some point in the future it might be possible to fix genes, perhaps before birth, to correct significant defects. DNA is also used to link suspects with evidence from crime scenes, and to demonstrate family relationships. Genome studies can tell us more about human evolution and our genetic links to other animals. The results of sequencing the chimpanzee genome were published in 2005. The 1,000 Genomes project, which aims to sequence the DNA of 1,000 individuals, aims to show the range of human genetic variety.

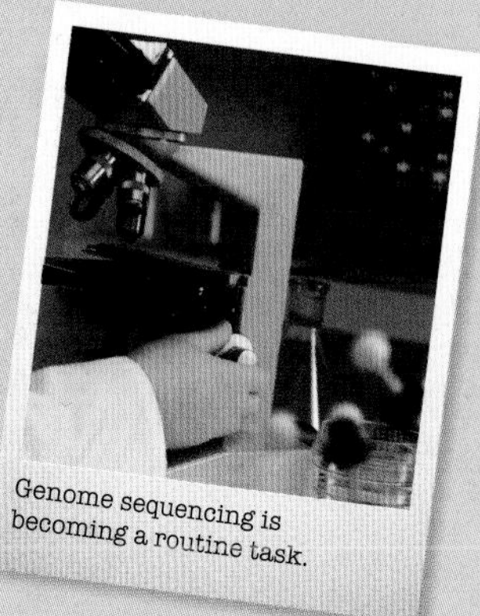
Genome sequencing is becoming a routine task.

1991
THE WORLD WIDE WEB

'A 'web' of notes with links between them is far more useful than a fixed hierarchical system.'

– Tim Berners-Lee, in his proposal for the World Wide Web

An archived copy of the world's first web page.

It's difficult to imagine life without the World Wide Web now, yet it's been with us for only around 25 years, and in its early years it was used by relatively few people. The web was developed at CERN, the European Organization for Nuclear Research, in Switzerland, by Tim Berners-Lee. Frustrated at the difficulty of sharing documents with colleagues working in the same area, he proposed a system for computers to share text and photographs with links between them, so that people could easily follow cross-references. He wasn't given permission to develop it as a CERN project, but pursued it on his own.

The early web was used first within CERN, and then with other research establishments and universities. Information kept on one computer used hypertext links to call up information kept on another computer, the two being linked by the internet. The first web page went live in 1991; it had links to the other 25 pages on the web. There are now more than five billion web pages.

Initially, the web was used only at academic and other research institutions. It became more widely accessible as better web browsers and search engines were developed. The first web browser to catch the public imagination was Mosaic, launched in 1993. It was the first browser to display text and pictures on the same page, rather then having all images open in a separate window. It was compatible with PCs, which by 1993 dominated home and business markets. 1993 also saw the first precursor of a search engine, called W3Catalog. It was not a real search engine, but a catalogue of the websites available. One of the first search engines that 'crawled' the web, building a database of content that could be properly searched, was WebCrawler, released in 1994. Search engines quickly proliferated. With a way of displaying text and images together, and a way to search for content, the web's popularity began its meteoric rise.

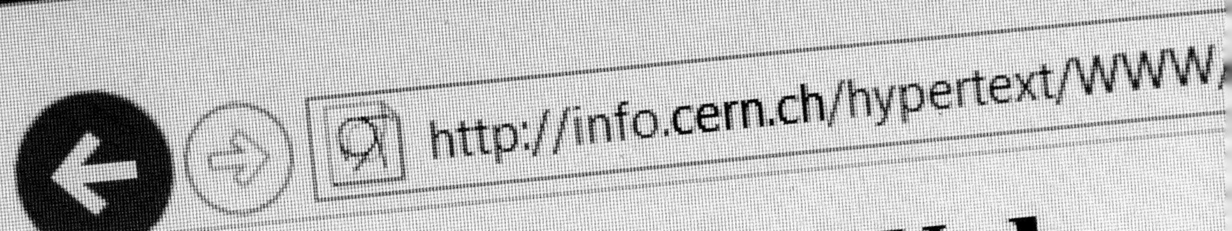

FLASHPOINT FACT

Tim Berners-Lee's boss responded to his response for funding for developing the web by saying his idea was 'vague, but interesting'. He was allowed to develop it in his own time, but not as an official project.

TIMELINE

1945 Vannevar Bush publishes his ideas for a machine that presents information in a linked form, and predicts an 'information explosion' later in the century.

1963 ASCII code is developed as a way of representing text that all computers can interpret.

1969 ARPANET links two computers in the USA.

1971 Ray Tomlinson sends the first email.

1973 Norway and England gain links to ARPANET, the first outside the USA to connect to it.

1973 FTP is implemented, allowing file transfer over the network.

1974 The term 'internet' is first used.

1978 Ward Christensen develops CBBS, a bulletin-board system to share news online.

1980 The Usenet news-sharing system is set up, presenting news and comments in threads.

FLASHPOINTS

1967
BUILDING THE NET

The World Wide Web is built on the internet, which grew from ARPANET. ARPANET was first proposed in 1967, and developed by the Advanced Research Projects Agency (ARPA) from US defence funding. In October 1969, ARPANET began by linking two computers in California – one at UCLA and the other at SRI – allowing researchers to share information long distance. It used a packet-switching protocol, which means that instead of a dedicated connection (like a phone line), communications are chopped into chunks, or packets, that are sent over shared connections and put back together at their destination. ARPANET soon expanded its reach, with other universities and research facilities linking in to the network.

1973
FROM ARPANET TO INTERNET

In 1973, the first computers outside the USA linked to ARPANET, beginning with computers in Norway and then England. ARPANET was not the only network. More networks grew up, both in the USA and elsewhere. In 1981, access to ARPANET was expanded and the following year the TCP/IP protocol provided a standardized method for computers to communicate. The concept of a worldwide network of computers all communicating with each other became fully realizable. Other networks linked to ARPANET, and in 1983 the internet was born, with ARPANET now just a part of it. The emergence of commercial internet providers during the 1990s made it fully available to commercial enterprises and even individuals.

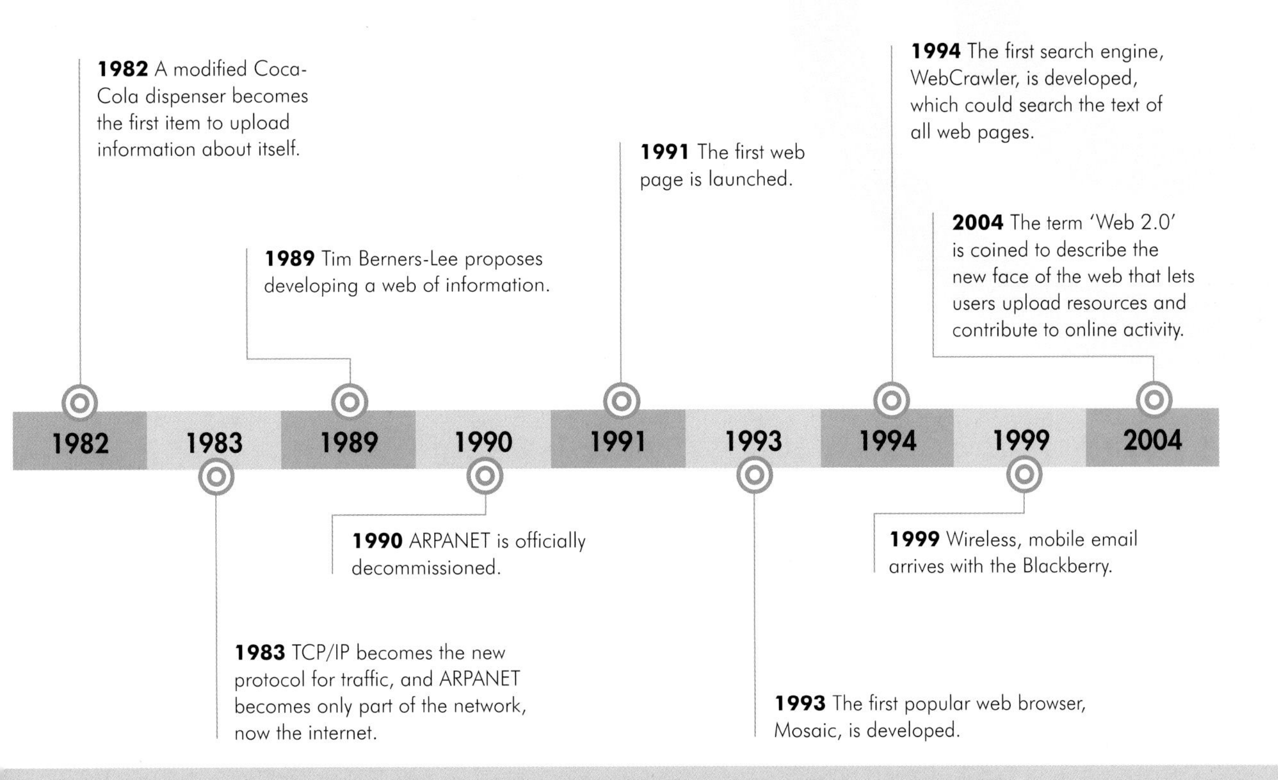

1982
IOT

The start of the next phase of the web has begun, the 'Internet of Things'. More and more devices are connected to the internet, but so far most of these have had human users behind them. The Internet of Things involves objects that have direct access to and interaction with the internet on their own, through automation rather than a personal choice to connect. The first internet 'thing' was a modified Coca Cola dispenser that, in 1982, could upload its inventory and show whether newly loaded drinks were cold. Wireless connections and miniaturization have made it possible for more things to be internet-connected. Predictions suggest that 20–50 billion 'things' will be connected by 2020.

Post 1993
FROM WEB TO WEB 2.0

The web was slow to take off with the public until the web browser Mosaic (1993) and early search engines made it easy to find and view web pages. Use of the web then grew steadily over the course of the 1990s. In 1994, there were 2,738 websites and around 45 million internet users; in 2000, there were over 17 million websites and over 400 million internet users. As the web grew, its focus began to change. In the early years of the 21st century, 'Web 2.0' began to emerge. This is a more dynamic interactive form of the web, in which ordinary users contribute content on social networking and file-sharing sites, and through blogs and wikis.

Social media is one of the most popular applications of Web 2.0.

1992

EXOPLANETS: HOMES FOR ALIENS?

'It is difficult to think of another enterprise that holds as much promise for the future of humanity.'

– Carl Sagan, of SETI

A NASA visualization of Kepler-186f, 490 light years from Earth, which might be habitable.

The first person we know to have believed that there are other worlds elsewhere in space was the Greek philosopher Democritus, who contemplated planets of different sizes 2,400 years ago. Interest in extra-terrestrial life flourished with the rise of science fiction from the 19th century, but only crystallized into a search for extra-terrestrial intelligence and the homes any aliens might occupy in the late 20th century.

The SETI (Search for Extra-Terrestrial Intelligence) project, set up in 1984 and operational from 1985, collects and analyzes radio signals from space in the hope of finding evidence of an alien civilization capable of electromagnetic transmission. At the same time, we are sending our own messages into space. The first exoplanet – a planet outside our solar system – was confirmed in 1992 when Aleksander Wolszczan and Dale Frail discovered two planets orbiting a neutron star. They were timing pulses of light from the spinning star and found anomalies that could only be explained by two planets passing in front of the star. They calculated in 2003 that each planet is around four times the mass of the Earth. In 1995, Michel Mayor and Didier Queloz discovered the first exoplanet around a regular star.

The confirmation of exoplanets has intensified interest in the search for extraterrestrial life, and as the total number of exoplanets continues to add up, scientists continue to search for evidence of another species in our universe.

FLASHPOINT FACT

In 2012, the Arecibo telescope beamed 10,000 Twitter messages into space, hoping to persuade aliens that this constitutes evidence of intelligent life on Earth.

TIMELINE

1877 | 1907 | 1959 | 1961 | 1972 | 1974 | 1977

1877 Giovanni Schiaparelli reported seeing networks of 'channels' on Mars, soon mistranslated as 'canals' and taken as evidence of life on Mars.

1907 Alfred Russel Wallace argues that Mars is not habitable.

1959 Giuseppe Cocconi and Philip Morrison suggest that we should be searching for radio waves from space that might be signals sent by aliens.

1961 Frank Drake publicizes his equation for calculating the probability of finding alien life.

1972 NASA *Pioneers 10* and *11* are launched, each carrying a plaque with an indication of their origins in case they are found by alien lifeforms.

1974 The Arecibo message is broadcast into space.

1977 The Wow! signal is identified in records of electromagnetic signals from space.

1961
DRAKE'S EQUATION

No one has any idea of how likely we are to encounter alien life forms. Is the universe teeming with life? Or are we perhaps the only intelligent life form in the universe at the moment? Frank Drake devised an equation in 1961 that showed how the probability can be worked out. It multiplies individual probabilities such as the chance of a planet having conditions that can support life, and the chances of lifeforms developing a civilization. Although we can't fill in many of the variables with much confidence, the equation has been hugely influential and has led to estimates ranging from one or two other civilizations to tens of thousands.

1974
WE ARE HERE

We have not just been waiting for aliens to contact us. In 1974, the 'Arecibo message' was broadcast, targeting the globular star cluster M13, 21,000 light years away. It includes pictures of the Arecibo telescope, a human, the structure of DNA and the solar system. As it would take 42,000 years to receive a reply, it was more a symbolic than a practical attempt at communication. When the *Pioneer* spacecraft were launched in 1972, and the *Voyagers* in 1977, both pairs of craft carried messages for other civilizations. The *Voyagers* carried Golden Records with sounds and pictures from Earth, and the *Pioneers* had a plaque showing where they came from and two human figures.

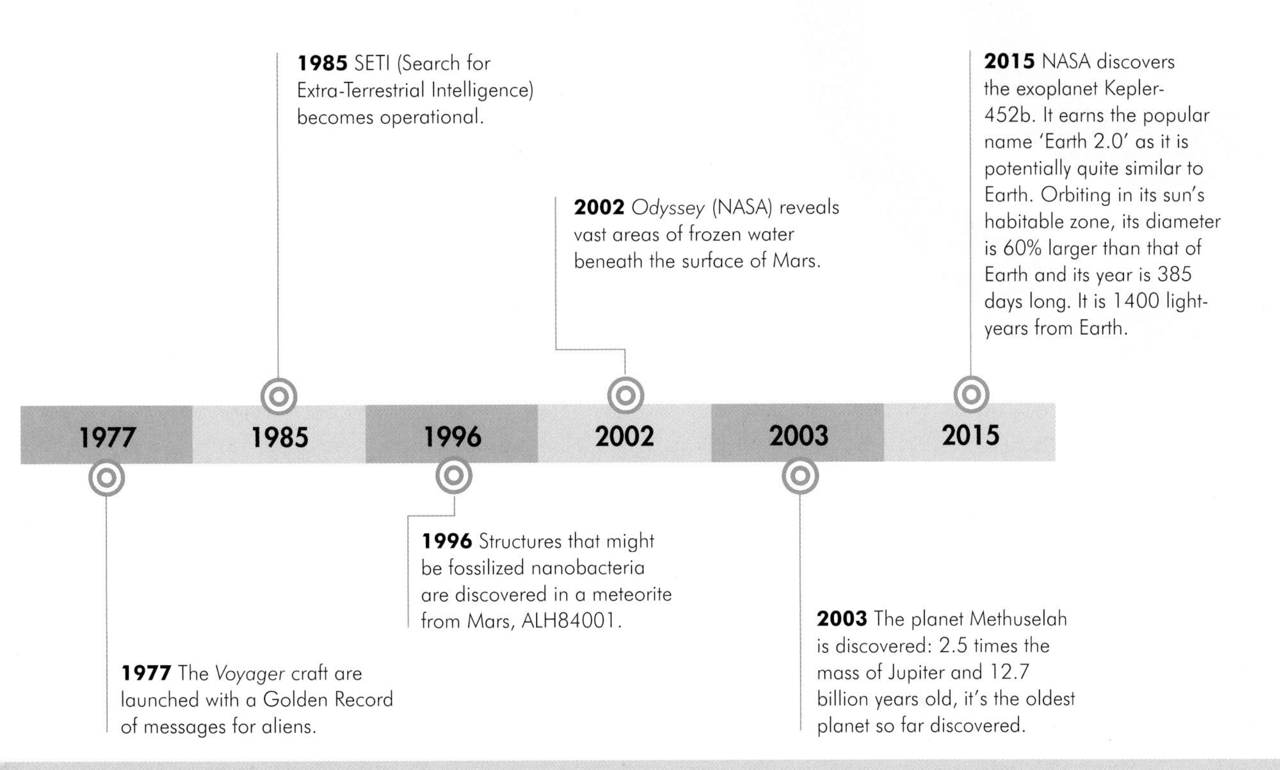

1977
LISTENING TO SPACE

In 1959, Giuseppe Cocconi and Philip Morrison suggested we should be examining electromagnetic radiation from space to look for communications from alien life. And so the hunt began. Nothing was found until, on 15 August 1977, Jerry Ehman spotted a signal with all the hallmarks of a deliberate transmission. It lasted 72 seconds and came from within the constellation of Sagittarius. As he labelled it 'Wow!' on the printout, it has become known as the Wow! signal. It has never been repeated, and no certain explanation of it has ever been found. The current SETI institute started work in 1985, and uses computers and crowd-sourcing to hunt for meaningful patterns in the reams of data it collects.

The Arecibo message, broadcast towards the star cluster M13 in 1974.

2002
ON OUR DOORSTEP

While radio telescopes search distant stars for signs of exoplanets, the search for life also continues closer to home, with investigation of the planets of our own solar system and their moons. In 2002, the NASA spaceship *Odyssey* found evidence of water ice in large quantities just below the surface of Mars. In 2011, researchers reported a vast ocean of water below the surface of Europa, one of Jupiter's moons. The presence of water or water ice is an indication that life is possible. It has now been found on many moons forming the rings around Saturn and in the body of comets.

The Very Large Array telescope in New Mexico, USA.

1998
IT'S ALL DARK

'Because dark energy makes up about 70 per cent of the content of the universe, it dominates over the matter content. That means dark energy will govern expansion and, ultimately, determine the fate of the universe.'

– Eric Linder, Director, Berkeley Center for Cosmological Physics

Is dark energy pushing the galaxies further apart at an ever-increasing speed?

The idea that we don't know much about the universe is not very new. But the idea that we actually only know anything at all about a tiny proportion is rather unsettling. The matter and energy we can see and measure in the universe seem to represent only five per cent of all there is. The rest – the vast majority – is accounted for by dark matter and dark energy.

Dark matter was first proposed in the 1930s to account for the apparent discrepancy between the mass of stars and galaxies and the speed at which they move. Dark energy was proposed later, in 1998, to explain the speed of expansion of the universe. Light from distant, exploding stars was found to be dimmer than is predicted by the redshift of light from their galaxies. Adam Riess, Saul Perlmutter and Brian Schmidt accounted for this by suggesting that around nine billion years ago, the speed of expansion of the universe increased. The universe is not only expanding, but the rate at which it is expanding is growing. There is no candidate for an energy source, so they suggested 'dark energy', which acts against gravity to push matter apart. Its nature is one of the biggest puzzles facing modern physics.

It is so puzzling that recently some scientists have started to question the notion of dark energy, wondering whether it actually exists at all. One argument against it is that it's based on the idea that the universe is homogenous – essentially the same in all directions. Yet we know that there is variety in the universe, with some areas more densely populated with matter and some showing great, empty voids. It might be, they suggest, that in the empty areas, space expands more rapidly than in the densely occupied areas where gravity counters expansion. But as Einstein demonstrated that time and space are linked, this would also mean that time 'expands' at different rates in different parts of the universe. The age of the universe would then depend on where you measured it – in a crowded place or an empty place. That idea might be even harder to accept than dark energy.

FLASHPOINT FACT

Most dark matter is not made up of the atoms and other particles that are familiar to us, does not carry an electrical charge and cannot interact with electromagnetic forces.

TIMELINE

1917 Albert Einstein introduces the cosmological constant.

1929 Edwin Hubble confirms that distant galaxies are moving away from us, and apparently moving faster the more distant they are.

1932 Jan Oort proposes dark matter to explain faster-than-expected movement of stars in the Milky Way.

1933 Fritz Zwicky suggests some unidentified dark matter adds to the mass of galaxies in the Coma cluster.

1970s Scientists come to agree that there is more to the universe than can be observed.

1978 Vera Rubin confirms the existence of dark matter.

1917	1929	1932	1933	1970s	1978

FLASHPOINTS

1917
SOMETHING CONSTANT

In 1917, Albert Einstein introduced the cosmological constant, represented by , for the energy density of a vacuum. He originally needed it to 'hold back' the effects of gravity and explain the steady state of the universe. Without the cosmological constant, the theory of relativity would give a universe that could expand or contract, which Einstein did not at the time believe happened. Now we are happy with the expanding universe and no longer need the cosmological constant for its original purpose, and from 1929 until the 1990s it was assumed to be zero. Then it came back into fashion in equations that attempt to explain dark energy.

1932
TOO FAST OR TOO LIGHT?

The first person to propose dark matter was Jan Oort in 1932, when he calculated that the stars in the Milky Way are moving faster than they should be, given what is known about gravity and their mass. Mass and velocity are closely related, and it should be possible to predict one from the other. The following year, Fritz Zwicky discovered that galaxies in the Coma cluster are also moving much more quickly than they should. His calculations suggested the galaxies had far more mass than previously thought. He proposed that they had a lot of 'dark matter' – matter that we can't see but that increases their mass. No one took much notice of this idea for 30 years.

The Milky Way.

1998 Adam Riess, Saul Perlmutter and Brian Schmidt suggest that dark energy is driving the expansion of the universe.

2003 The National Research Council lists 'What is the nature of dark energy?' as one of the most pressing scientific problems to be answered in the 21st century.

2012 The Dark Energy Camera starts operation in Chile, collecting light from galaxies eight billion light years away. It will take five years to collect data that might help astronomers understand more about dark energy.

2013 The relative proportions of normal matter, dark matter and dark energy are determined.

2013 The latest measurement of the acceleration of the universe's expansion sets it at 74.3km (46.1 miles) per second per megaparsec (a megaparsec is roughly three million light years).

2015 Construction of the LSST telescope starts in Chile. It will have the most powerful digital camera ever made and scientists hope it will shed some light on dark energy. It will become fully operational in 2022.

1978
PINNING DOWN DARK MATTER

In the 1960s, Vera Rubin and Kent Ford found that the Andromeda galaxy, too, confounds expectations. Using a new spectrographic tool, they measured the orbital speed of stars and gas at different distances from the centre of Andromeda. Those further out should orbit more slowly, but they travel just as fast as those near the centre. Rubin led a team to investigate dozens of galaxies. In 1978, she published a startling conclusion: that spiral galaxies are embedded in a much larger sphere of dark matter, far more massive than the visible galaxy. Dark matter does not emit or radiate electromagnetic radiation. Several theories and hypothetical particles have been suggested to explain dark matter; the most popular is WIMPs – weakly interacting massive particles.

The Andromeda galaxy (M31), 2.5 million light years away.

2013
WHAT IS THERE?

In 2013, the results were published of the Planck space observatory's monitoring of anisotropies in cosmic background radiation. This measurement of tiny differences in the background radiation, along with other data collected, enabled a more accurate dating of the universe, to 13.798±0.037 billion billion years. It also gave a calculation of the relative density of different types of matter and energy in the universe. The universe, it appears, contains around 4.82 per cent ordinary matter, 25.8 per cent dark matter and 69 per cent dark energy. Ordinary matter is the bit we are familiar with – leaving around 95 per cent of the universe still to be uncovered..

INDEX

Aldrin, Buzz 107, 136
aliens 157, 166–9
Allen, Paul 133
Altair 8800 132, 134
ammonia 22, 24–5, 71
Andromeda galaxy 173
antibiotics 44–7, 120
antibodies 114, 115
antimatter 126
Apollo missions 106, 107
Apple 132, 135
Arecibo telescope 63, 166, 168, 169
Aristotle 54
Armstrong, Neil 107
ARPANET 164, 165
artificial intelligence 157
Askanazy, Max 152
asteroids 138, 139
astronomy 52–5, 60–3
atom bombs 74–7
atomic structure 26–9, 37, 124–7

Babbage, Charles 134
background radiation 173
bacteria 47, 114, 120, 122
Baird, John Logie 40, 43
Bakelite 48, 50
Bardeen, John 82, 84
Barnard, Christiaan 93
Becquerel, Henri 76
Behaviourism 20
Berg, Paul 122
Berners-Lee, Tim 162, 163, 165
big bang theory 37
Big Bang theory 52–5, 156, 157
biotechnology 123
black-box radiation 6
black holes 154, 156, 157
Bletchley Park 64–5, 66, 134
blood cells 152
blood groups 10–13, 92
Bohr, Niels 6, 8, 28, 76
bone-marrow 152
Bosch, Carl 22
bosons 124, 126, 154
Brattain, Walter 82, 84
A Brief History of Time (Hawking) 154
Brown, Barnum 14–17
Brown, Louise 140, 143
Brownian motion 36, 37
Burgess Shale 17

calculators 134
carbon dioxide 116–19
Carothers, Wallace 48, 50, 51
Carson, Rachel 108, 110
cells 88
cellular respiration 68
CERN 127, 162
Chadwick, James 29, 76, 126
Chain, Ernst 44, 47
chain reactions 74–7
chemical reactions 70
Chernobyl 77
chlorine 25
chromosomes 88, 158, 161
Churchill, Winston 112
classical conditioning 18
climate change 116–19
clones 112–15
Cold War 73, 103
Collins, Francis 158
Colossus 134
comets 139
computers 64–7, 83, 85, 103, 132–5, 145, 162–5
conservation 108–11
continental drift 30, 33
cosmic microwave background radiation (CMBR) 55
cosmological constant 172
Crick, Francis 86, 89, 122
Curie, Marie 76
Cuvier, Georges 16
cyclosporine 93

Dalton, John 70
dark energy 170–3
dark matter 124, 170, 171, 173
Darwin, Charles 130
DDT 108, 109, 110, 111
De Broglie, Louis 29
Denys, Jean-Baptiste 12
diabetes 71
dinosaurs 14–17
diphtheria 94, 96
diseases 114, 161
DNA 86–9, 120–3, 158
dogs in space 106
Dolly the sheep 112, 113, 115
Domagk, Gerhard 46
Doppler shift 54
Drake's equation 168
Dubois, Eugène 130
dust bowl 110

earth, structure of 30–3
Earth Summit 119
earthquakes 33
Edwards, Robert 140, 143
Ehrlich, Paul R 116
Einstein, Albert 6, 8, 28, 34–7, 52, 76, 124, 156, 172
electromagnetic spectrum 40, 42
electromagnetism 42, 124
electron microscopes 56–9
electrons 8, 9, 26, 28, 58, 124, 126
email 164
environmental scanning electron microscope (ESEM) 56
Evans, Martin 150
event horizons 156
evolution 17
exoplanets 166–9
Experimental Breeder Reactor 77
Explorer 100
explosives 24

Fermi, Enrico 74
fertilizers 22, 24
First World War 13, 24
Fleming, Sir Alexander 44, 46
Florey, Howard 44, 47
Ford, Kent 173
fossils 14–17, 128–31
Franklin, Rosalind 122
Friedmann, Alexander 52, 54
Fukushima Daiichi nuclear power station 77

Gagarin, Yuri 104
Gaia hypothesis 111
galaxies 52, 54
Galileo 62
Gates, Bill 133, 149
Gell-Mann, Murray 124, 126
genes 86–9, 115, 158–61
genetic diseases 161
genetic engineering 120–3
global warming 111
glucose 68
gluons 126
GM foods 122, 123
Goddard, Robert 78, 80
Golden Record 136, 169
Gore-Tex 51
GPS 101
graphical user interface 135
gravity 9, 34, 37, 156, 172
greenhouse effect 116, 118, 119
Gurdon, John 112, 114

Haber-Bosch process 22, 24–5
Hadron Collider 127
Hawking, Stephen 150, 156, 157
Herrick, Ronald 90
Hertz, Heinrich 42
Higgs, Peter 126, 127, 154
Hiroshima 74, 76
Hodgkin, Dorothy 47
Homo erectus 130
Hoyle, Fred 52
Hubble, Edwin 37, 52, 54, 172
Hubble Space Telescope 63, 101, 157
human ancestry 128–31
human body 68–71
human genome project 158–61
Huntington's chorea 161
Huygens, Christian 8
hydrogen 22

IBM 132
infinity 54
infrared 42
inheritance 86, 88
insulin 71, 120, 123

integrated circuits 85
International Space Station 107
internet 162–5
Internet of Things 165
isotopes 74, 76
IVF 140–2

Jansky, Karl 60, 61, 62
Jenner, Edward 96, 148
Jobs, Steve 135
Jodrell Bank 62
Johanson, Donald 128
Jupiter 138, 139, 169

Kaufman, Matthew 150
Kennedy, John F. 106
Knoll, Max 56, 58
Koch, Robert 96
Krebs cycle 68–71
Kyoto Protocol 119

Lamb, Willis 124
Landsteiner, Karl 10–13
Lavoisier, Antoine 70
Lemaître, Georges 52, 54, 156
leukaemia 153
light 6–9, 29, 36, 42
liver 71
Lorenz, Konrad 21
Lovelock, James 111, 119
LSST telescope 73
Lucy 128–31

magma 32, 33
Manabe, Syukuru 116, 117, 118
Manhattan Project 36, 74, 76
Marconi, Guglielmo 40, 42, 43
Mars 136, 138, 139, 168, 169
Maxwell, James Clerk 8, 42
medicines 114
Mendel, Gregor 88
Mercury 139
metabolic pathways 68–71
Methuselah 169
microbes 96
microbiology 56–9
microscopes 56–9
Microsoft 132, 133, 135
Miescher, Friedrich 88
mitochondria 68
Mlilikan, Robert 8, 9, 36
MMR 97
mobile phones 43, 83
molecules 50
moon landings 106, 107

Morris, Desmond 130
Morse code 42
MRSA 47

Nagasaki 76
The Naked Ape (Morris) 130
NASA 78, 81, 100
nebulae 54
neoprene 48, 50
neutrons 26, 29, 74, 76, 124, 126
Newton, Isaac 8, 42
nitrates 22–5
nitrogen 22–5, 70
nuclear bombs 74–7
nuclear power 77
nuclear reactions 29, 36, 124
nylon 48

Oberth, Hermann 78, 80, 81
Odyssey 169
Oort, Jan 172

Pangaea 30
particle zoo 126
Pascal, Blaise 134
Pasteur, Louis 46, 70, 71, 96
Pauli exclusion principle 9
Pavlov, Ivan 18–21
penicillin 44, 46, 47
pesticides 108, 110
phages 46, 122, 160
pharming 115
phobias 20
photoelectric effect 36
photons 8, 9, 28
Pioneer missions 138, 139, 168, 169
Planck, Max 6–9, 36
plastic bags 51
polio 94–7
pollution 108–11
polyethylene 51
polymers 48–51
polythene 50
positrons 126
printed circuits 84
prontosil 46
protons 26, 124, 126
psychology 18–21
pulsars 62, 63
punched cards 134

quantum theory 6–9, 28, 36
quarks 124, 126, 127
quasars 63

radar 42, 43, 51

radio 40–3
radio astronomy 60–3
radioactivity 76, 124, 152
radios 84, 85
Reber, Grote 60
reflexes 18
relativity 34–7
respiration 68–71
Richter scale 32
RNA 89, 122
Röntgen, Wilhelm 42, 76
Roosevelt, Franklin D. 140
Rosetta 139
Rubin, Vera 173
Ruska, Ernst 56, 58
Rutherford, Daniel 24
Rutherford, Ernest 26, 29, 76

Sabin, Albert 94, 97
Sagan, Carl 166
Salk, Jonas 94, 97
Saltpetre War 24
Sanger, Fred 160
satellites 98–101
Saturn 138
scanning electron microscope 56, 59
sea-floor spreading 33
sea urchins 151
Second World War 43, 44, 73
seismographs 32
semi-conductors 84
SETI 63, 166, 168, 169
Shockley, William 82, 84
silk 48
skin grafts 92
Skinner box 21
Skylab 107
smallpox 97, 146–9
smog 111
solar energy 36
solar system 136–9
space 166–9
space race 98–101, 103, 104–7
space rockets 78–81
space shuttles 107
space stations 101
space-time 34, 55, 170
Spallanzani, Lazzaro 142
special relativity 34
speed of light 34, 37, 40
Spektr-R 63
Spemann, Hans 114
sperm 142, 143
Sputnik 78, 81, 98–101, 106, 138
stem cells 93, 150–3

Steptoe, Patrick 140, 143
string theory 127
sulpha drugs 46, 47
superstring theory 127
Szilárd, Leó 74–7

T. Rex 14–17
Taung Boy 128, 130
Taylor, Geoffrey 8
tectonic theory 33
Teflon 48, 49, 50, 51
telegraph 42
telescopes 60–3
television 40, 43
Tereshkova, Valentina 106
test-tube babies 140–3
Thompson, J.J. 26, 28, 58, 126
Three Mile Island 77
3D printers 153
time dilation 37
transfusions 10–13
transistors 43, 82–5
transplants 90–3, 114, 152
Turing, Alan 64–7
Turley, Jim 82

ultra violet 42
Uncertainty Principle 9
urea cycle 71
Ussher, Archbishop 32

V-2 rockets 81
vaccines 94–7, 146–9
vacuum tubes 84
Venera missions 138
Venus 138
viruses 46, 56, 59, 114, 122, 149
Von Braun, Wernher 78, 81
Von Leeuwenhoek, Antonie 58
Voyager missions 136, 169
vulcanization 50

Watson, James 86, 122
Watson-Watt, Robert 43
wavelength 7, 40
Wegener, Alfred 30–3
WIMPs 173
World Health Organization 146, 149
World Wide Web 135, 162–5
Wowl signal 168, 169
Wundt, Wilhelm 20

X-rays 9, 42, 76, 86

Zwicky, Fritz 172

PICTURE CREDITS

Alamy BSIP SA 54; Capt.digby 42; Dennis Kunkel/Microscopy/Phototake 59, 122; Historic Florida 51a; Interfoto 21, 46, 83, 134, 135; ITAR-TASS 67; Karen Fuller 65; Mark Dunn 66; Pictorial Press 19, 39l; RIA Novosti 102, 106; Robert Preston Photography 129; Sebastian Pionka 77; The Natural History Museum, London 130; World History Archive 87

Brigham and Women's Hospital Archives 91

Corbis Bettmann 11, 15, 53, 105, 11r, 147; David Scharf 151; George W Wright 111; Jim Sugar 155; Karen Kasmauski 113; Marli Miller/Visuals Unlimited 17; Paul McErlane/Reuters 144; Paul Souders 117; Underwood & Underwood 109

Getty Images AFP/Fabrice Coffrini 163, /Jacques Demarthon 131; Bloomberg 115; Boston Globe 90; Cincinnati Museum Center 95; David McNew 153; Fred Ramage 57; goktugg 5l; Hulton Archive 49, 79, 80, 141; Mondadori 73r; Oxford Science Archive/Print Collector 31; PhotoQuest 41; Roger Viollet 5r; SSPL 4, 27, 35, 35, 37, 38, 47, 50, 72, 81, 123; Thomas D Mcavoy/Life Picture Collection 61, 63a; Ullstein bild 39r; Universal Images Group 16, 63b

iStockphoto.com Kazakovmaksim 112; Kiyoshi Takahase Segundo 143

NASA 55, 100, 107, 137, 138, 139, 156, 167, 173, ESA/S Beckwith and the HUDF Team 157; Frank Drake (UCSC) et al Arecibo Observatory (Cornell, NAIC) 169l

Press Association Images William Berry/AP 133

Science Photo Library A Barrington Brown 73l; American Institute of Physics/Emilio Segre Visual Archives 23, /Margarethe Bohr Collection 7; Arscimed 126; British Antarctic Survey 119; Carlos Clarivan 125; Claus Lunau 29l; Dr Gopal Murti 121; Eye of Science 149; I Curie & F Joliot 29r; Jose Antonio Penas 28; NYPL/Science Source 92; Omikron 26; St Mary's Hospital Medical School 45

Shutterstock Sergey Panychev 84; 18percentgrey 161; agsandrew 8; Albert Barr 171; Ashwin 43; chromatos 71; daulon 33; decade3d - anatomy online 96; Dmitry Kalinovsky 93; Everett Historical 75, 110l; fotokdravat 172; Gio.tto 145r, 159; Hannahmariah 97; ivannn 25b; Johan Swanepoel 32r; Jubal Harshaw 114; kanvag 51b; Ken Schulze 36; Kenny Tong 169r; mama_mia 145l, 165r; Marioner 69; molekuul.be 70; The Goatman 85; www.royaltystockphoto.com 152

Thinkstock Cathy Yeulet 142; Dorling Kindersley 24; HotPhotoPie 103r; Ian Bracegirdle 118; Photos.com 12, 62; TopFoto Heritage Images 98, 99

US National Library of Medicine via Wikipedia 20

Wikipedia/Creative Commons 32l, 58, 88, 103l, 127, fdardel 160.